Also by Michael Lewrick and Larry Leifer:

The Design Thinking Toolbox by Michael Lewrick, Larry Leifer, and Patrick Link

The Design Thinking Playbook by Michael Lewrick, Larry Leifer, and Patrick Link

THE DESIGN THINKING LIFE PLAYBOOK

THE DESIGN THINKING LIFE PLAYBOOK

EMPOWER YOURSELF, EMBRACE CHANGE, AND VISUALIZE A JOYFUL LIFE

MICHAEL LEWRICK
JEAN-PAUL THOMMEN
LARRY LEIFER

DESIGN
ACHIM SCHMIDT

WILEY

Originally published by Versus Verlag AG.

Published by John Wiley & Sons, Inc., Hoboken, New Jersey.

Published simultaneously in Canada.

For general information on our other products and services or for technical support, please contact our Customer Care Department within the United States at (800) 762-2974, outside the United States at (317) 572-3993 or fax (317) 572-4002.

Wiley publishes in a variety of print and electronic formats and by print-on-demand. Some material included with standard print versions of this book may not be included in e-books or in print-on-demand. If this book refers to media such as a CD or DVD that is not included in the version you purchased, you may download this material at http://booksupport.wiley.com. For more information about Wiley products, visit www.wiley.com.

ISBN 9781119682240 (Paperback)
ISBN 9781119682257 (ePDF)
ISBN 9781119730439 (ePub)

Cover image and illustrations: Achim Schmidt

V2D93891A-DF1E-4A7C-B840-0C3CB71A73C8_070720

This is the DTL Playbook of:

. . . .

About the authors:

Michael Lewrick, PhD, is a featured speaker and teaches Design Thinking at various Universities. He is the author of *The Design Tthinking Toolbox* and the international bestseller *The Design Thinking Playbook*, in which he describes the mindful transformation of people, teams, and organizations. Michael is a thought leader in applying different mindsets to solve thorny problems. He collaborates with colleges, universities, and companies and focuses on the self-efficacy of people in personal and organizational change. He is an internationally recognized leader in the field of digitization, innovation, and the management of change.

Jean-Paul Thommen, PhD, has coached and developed students and managers for many years. At the European Business School (Wiesbaden), he established a coaching program that thousands of interested people have successfully completed over the last decade. His German publication *Coaching* is one of the foundational works in this field. He is a professor lecturing at various colleges and universities in the fields of leadership, organizational development, and business ethics. He has also advised large and small enterprises on these topics.

Larry Leifer, PhD, is one of the most influential personalities and pioneers of design thinking in the world. He has introduced design thinking globally and guided numerous companies, innovation practitioners, and student teams in their search for new market opportunities. Over the years, he has developed various design thinking techniques and adapted them to the individual needs of the design thinking life. Furthermore, he is a professor of engineering design and the founding director of the Center for Design Research at Stanford (CDR) and the Hasso Plattner Design Thinking Research Program at Stanford.

Contents

Why we're taking you along with us on this journey	8
Foreword	10
Introduction	13
Part I: Apply the DTL mindset	24
Reflect, accept, and understand	42
Keep an energy journal	56
Integrating other people's observations and perceptions of us	88
Define point of view	98
Find and select ideas	110
Design, test, and implement life plans	146
Self-check	178
Part II: Professional and career planning	184
Design career paths	214
Evaluate, test, and implement options	226
Questions for reflection on options	244
Last but not least – The journey's end is really the beginning	248
Index	253

Why we're taking you along with us on this journey

Because empathy and self-efficacy are essential for further development.

Because you will have moments when you wonder what comes next.

Because your life is one of the biggest and most complex projects you will ever undertake.

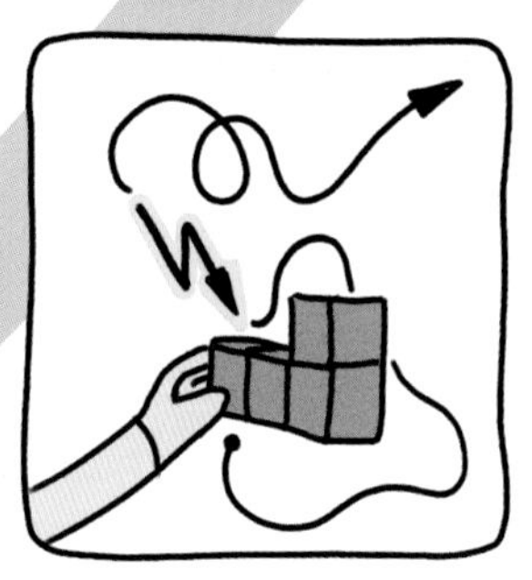

And, above all, because you can initiate a change in yourself at any time.

Because design thinking is a holistic approach that provides you with the tools to dig deeper.

Because a positive mindset helps you experience more happy moments in your life.

Because another way of looking at problems helps you bring about change.

Because a personally efficacious life is a happy life.

Because a life that you can visualize can be better realized.

Foreword

Bernie Roth

- Professor of mechanical engineering
- Co-founder of the Hasso Plattner Institute of Design at Stanford (the d.school)
- Author of *The Achievement Habit*

I have had the pleasure of teaching design courses for more than 60 years. Originally, I mainly taught about designing ordinary machines. Then I began to include designing robots. From there, I expanded into fostering creative thinking and improving students' personal efficacy. About 15 years ago, I was one of the founders of the Stanford d.school, and that transitioned my main activities from design to what we now call design thinking. Design thinking is a range of methods that apply techniques and ideas, originally used primarily to design physical objects, to problems dealing with a much wider range of issues.

At the Stanford d.school, the original bias in design thinking was toward designing for others. It was felt that objective outsiders brought insights that someone immersed in his or her own issues might overlook. We developed the mantra "Don't design for yourself!" This created a conflict for me because by then a lot of my teaching had to do with getting individuals to take control of their own lives by applying the very same principles that became part of design thinking.

So, when I started to write my book *The Achievement Habit*, I was concerned that my design thinking colleagues would accuse me of heresy. Instead, the opposite happened; it became acceptable to apply design thinking to oneself. For example, Tim Brown, then CEO of IDEO (one of the world's leading design thinking consultancies), wrote on the book jacket of *The Achievement Habit*: "Before unleashing design thinking on others, unleash it on yourself. You, and the world, will be far better for it." Even more gratifying to me were the e-mails from numerous readers telling me how grateful they were for the positive changes they were bringing into their lives.

In the last few years the design thinking world has evolved, and there have been other books guiding readers to apply design thinking to their own lives. This makes perfect sense because one of the most important principles in design thinking is to be user-centered. This implies empathy by the designer for the user. Empathy means the designer seeks to "step into the shoes" of the person for whom something is being designed. So, clearly, if people are designing for themselves, they are already in the shoes of the person they are designing for. However, there is a danger because we tend to be less objective when it comes to self-assessment. It is therefore important that the problem-solving methods we apply to our own lives include safeguards to alert us when we are deluding ourselves or simply not seeing things clearly. I was pleased to note that the authors of *The Design Thinking Life Playbook* are aware of these potential pitfalls, and they do a good job of alerting their readers. This is one of the strong points that makes this book a welcome addition to the field of design thinking.

The methods set forth in *The Design Thinking Life Playbook* present tools for people who seek to make their lives more fulfilling and have the courage to look at their current situation honestly. The techniques and strategies that Michael Lewrick, Jean-Paul Thommen, and Larry Leifer lead their readers through in this workbook can be used to redesign one's life, in terms of both one's activities and one's relationships. Going through the exercises should prove to be a valuable experience for all who want to initiate change and have the courage to think, act, and take advantage of life's opportunities.

–Bernie Roth

Welcome from the authors

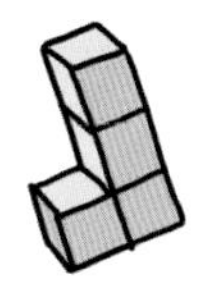

In this *DESIGN THINKING LIFE (DTL) Playbook*, we would like to present a selection of techniques and strategies that help initiate change. These tools work best when they are adapted to the specific situation and when people in our environment also know which "journey" we are on right now. This is why it makes sense that our circle of friends, family, and colleagues also have the *DTL Playbook* at hand and embark on this journey themselves or together with us. When dealing with DTL, we will quickly notice that change also has a lot to do with our social environment. The question of our self-image versus Other people's perceptions of us is an integral part of our social system.

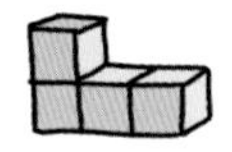

As in design thinking, the DTL process presented is used for orientation, i.e., to know where we are in the DTL cycle. A flexible application, adapted to the situation and the respective topic, is critical for success. Finally, the effect and the effectiveness are critical.

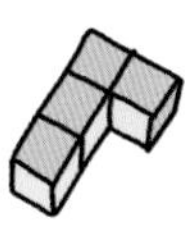

This playbook deliberately dispenses with scientific treatises on the topic of life design and puts application at the heart of things. Of course, it is desirable to apply any kind of new knowledge – for example, from the neurosciences – to one's personal life, to deepen one's understanding of the subject by reading specialist literature, or to adapt the techniques presented.

It is also very important for us to point out at the beginning of this playbook that certain topics are extremely complex and our emotions prevent us from helping ourselves. When we notice such signs, it is important to take advantage of professional coaching – with an appropriate expert. Professional coaches don't bite!

–Michael Lewrick, Jean-Paul Thommen, and Larry Leifer

Introduction

How is the DTL Playbook structured?

We have deliberately divided the book into two parts.

In the first part, we accompany Sue and John in their desire for change. We present strategies and techniques that help you act in a self-efficacious way. Typical questions are:

- What do I enjoy?
- What robs me of energy, and how do I recharge my batteries?
- What small changes can I initiate on my own and experiment with?
- How can I implement these changes?

The second part deals with the big decisions and changes in life. The focus is deliberately on professional and career planning because we have discovered that it is an issue we feel confronted with at regular intervals, from school age to retirement. We accompany Steve as he questions himself about career planning. Typical questions are:

- What are my skills and talents?
- Where can I apply them profitably?
- How do I find out what I like in order to make the right choice in terms of course of study, profession, and career?
- How do I choose between different options?
- How do I prepare myself to leave my comfort zone and initiate a change?

What is the best way to use this playbook?

The *DTL Playbook* provides us with different strategies and techniques to start the process of change. We recommend that you study Part 1 of the book first and Part 2 as a second step if the topic of career is currently vital to you.

DESIGN THINKING LIFE!

It is never too late to initiate a change.

The strategies, techniques, and examples presented are aids, which we adapt according to the situation.

- We follow the DTL process presented.
- We implement the instructions and fill in the empty pages with our thoughts.
- We note down our wishes, sketch out our milestones, and evaluate our experiences.
- We find a way to change based on our own needs.
- Achieving self-efficacy should become our top working motto.
- We take time for this journey on which we learn more about ourselves, try out new things, and gradually introduce change.
- We use the examples of John, Sue, and Steve as inspiration and guidance in implementing the tools presented.

Who are John, Sue, and Steve?

John, Sue, and Steve are people like you and me. They have arrived – as we might have – at a point in life where they would like to change something. The three fictitious characters (so-called personas) work, at different points in the DTL book, on their new milestones in life and thus help us with the application of the tools presented.

A new stage in **John's** life is just beginning. After a successful career in the field of service, he is going into early retirement at the age of 55. His children have already left home. He now has time for his hobbies and for his great passion, riding his motorcycle.

But this new phase of life also has its pitfalls; his relationship with his wife, for example, could do with some new ideas. John also wants to design this phase of his life actively and uses a number of DTL tools to do so.

Our second persona is **Sue**; she's in her mid-30s. After her studies and international career in marketing, she realizes that something is missing in her life.

Sue wants a life partner. She also misses her parents and siblings who live in Switzerland. Her life and work in Hong Kong are exciting and challenging, but not fulfilling in the long run.

Steve is still at the beginning of his professional career. He has just graduated from Stanford University with a bachelor's degree in business information systems and is uncertain whether to pursue a master's degree or accept a job in a start-up.

The start-up sounds exciting, but Steve has no work experience at all. Steve's role model has always been his older brother Alex. After high school, Alex stayed and lived close to his parents in upstate New York; immediately after getting his bachelor's degree, he earned his master's degree. Now he is in the final phases of earning a doctoral degree at Cornell University. His big dream is to live and work in Singapore. However, Steve also finds this path very tedious and lengthy.

We return to Steve and Alex, who are both facing major life changes, in the second part of the DTL book. We will explain later why this happens only in the second part and why we start with small changes.

The three fictional characters are the result of our DTL work at companies, at universities, and in countless coaching sessions. All three characters have initiated a change and have steered their lives in a new direction. For them, DTL has become a continuous process of reflection, self-efficacy, and adjustment that is never completed. DTL has become the basis for the creation of a satisfied and happy life.

For all of us, the individual design of a fulfilling life is moving more and more into the center, because the world we live in is becoming more and more difficult to orientate and demands that we commit to high performance. The growing complexity and high performance demands are difficult to keep at bay, so we have no choice but to develop strategies to deal with them in the best possible way. Reality is exciting, and you can write your own personal scenario for life if you want to!

> "If your mindset is unprejudiced...it is open to everything. 'In the beginner's mind, there are many possibilities, but in the expert's mind there are few.'"
>
> – Shunryu Suzuki

What is design thinking?

In design thinking, we adapt methods that are commonly applied by designers. This is why we make use of an iterative approach in design thinking, from the problem statement right up to a problem solution. Supported by various creativity techniques, the aim is to generate as many and at times "wild" ideas as possible. The creative working method aims at triggering both halves of our brain. On our "journey" to a solution, iterations, leaps of imagination, and combinations of ideas are desirable so that ultimately we arrive at a solution that meets the needs of people. On the way to the solution, a high level of error tolerance is of great value, particularly in the early phase. The techniques and strategies presented in this book are a means to an end; that is, you always adapt the tools to your situation.

A pivotal aspect of the design thinking mindset is to free yourself from prejudices and assumptions. This means to be open to a world of possibilities because at the beginning of the "journey," we do not yet know what is possible and what is not.

In design thinking, we use a persona, a fictitious character that has certain needs and for whom we work out a solution. In the DTL book, we also have personas but with the aim of showing how a fictitious person solves a problem. The respective solutions are only examples of an individual change. **They are not sample solutions or the authors' recommendations for your life!**

The design thinking process and the mindset we use in the DTL book will be discussed in more detail later.

The design thinking mindset means:

- We **say goodbye to prejudices** about **"how things work."**
- We **put aside expectations** about what will happen.
- We **are curious** so as to **understand facts and problems in depth.**
- We **open** ourselves up to **new possibilities.**
- We **ask simple questions.**

If you want to know more about design thinking, refer to:

- *The Design Thinking Playbook* by Michael Lewrick, Patrick Link, and Larry Leifer
- *The Design Thinking Toolbox* by Michael Lewrick, Patrick Link, and Larry Leifer

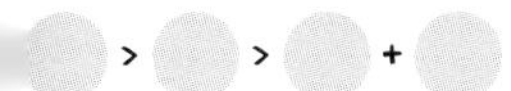

60-minute warm-up "Back to the Future"

Imagine you can embark on a time travel journey with Marty McFly and Doc Brown from the successful film trilogy *Back to the Future*. Your TIMER is geared to the future; exactly ten years from today.

TEMPORAL DESTINATION

1) Start the journey!

10 min

The journey begins. There is a short flash, blinding light, and you arrive in the future version of your life. As we know from *Back to the Future*, there are sometimes cracks in the space-time continuum, and you may be teleported to a different location in the future.

Make a sketch of what you see in this flash-forward.

You think you can't sketch?
Use these sketching shorthand techniques or devise your own.

Shapes | People | Arrows | Faces

2) Reflect on your journey through the space-time continuum!

a) What is the key feature of your vision of the future of your life?

5 min

b) Who were the people in your environment? Were they similar to people today or different?

5 min

3) Write a headline in a newspaper of the future!

10 min

THE TIMES
PhD with HONORS
Graduate offered CEO position at Google
NEWS

Write the headline and the first two sentences of the story of your life. Don't just describe the situation, but, above all, emphasize the news.

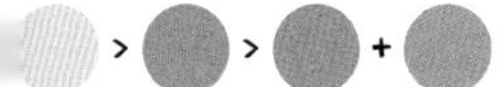

4) Act it out!

10 min

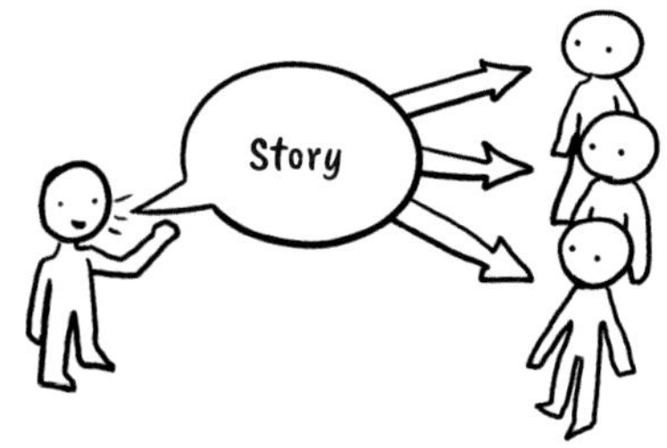

Tell someone from your circle of friends, family, and acquaintances about your journey into the future.

Show your vision of the future to a third person and ask what he/she thinks.

5) Create a timeline!

10 min

Think about four stages, starting today, that will lead you to your vision in ten years.
Draw the stages on a timeline.

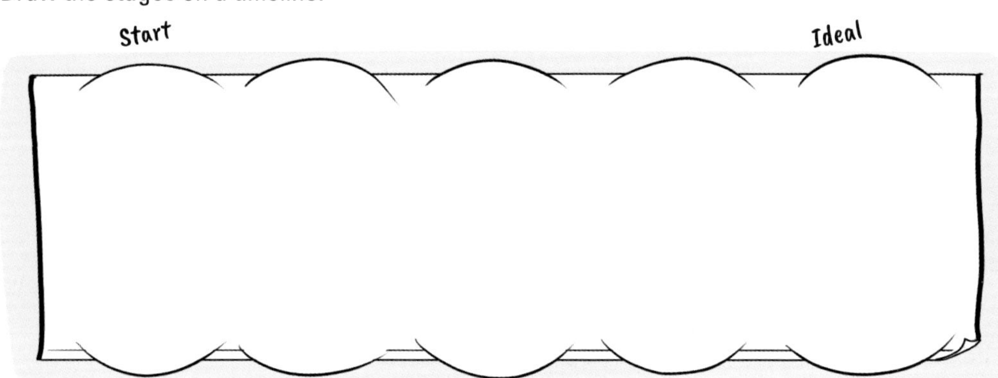

6) Framework conditions

5 min

Think of three conditions that need to change, so that you could tackle the first stage.

7) Invest fifteen minutes now!

15 min

Do something that initiates the change. For example, put together a to-do list or browse the Internet to find out more about what you would like to do or just go for a dream walk, during which you imagine what your future will be.

I'm allowed to
& I will

Some of these actions will help us get ahead; others are a dead end. But every change does something positive. Use discarded life plans to learn from them and design your future!

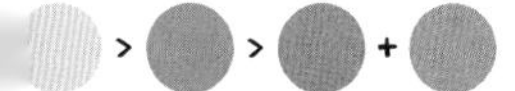

Part I

Apply the DTL mindset

In our warm-up "Back to the Future," we have already imagined a possible future in our minds. But "DESIGN THINKING LIFE" is about more: Above all, we want less stress, more well-being, and greater satisfaction. And who among us doesn't want that? Many people are dissatisfied with their current lives: at work, in their relationships, or with their personal circumstances. Reason enough to change something about it and become active yourself? If not now, then when? The "DESIGN THINKING LIFE" mindset helps you design your future consciously.

Which mindset helps us initiate change?

The design thinking mindset actively initiates positive change and transformation. So it is high time to apply this mindset to the topic of life design – now!

To change our lives, we need some courage, the ability to reflect on ourselves, the willingness to criticize ourselves, and finally a personal vision of the change. The "DESIGN THINKING LIFE" mindset gives us techniques that help us develop and enhance our own lives. These techniques include the exploration of our own needs, the search for new ideas, and the willingness to try something new before we initiate the change in repetitive steps.

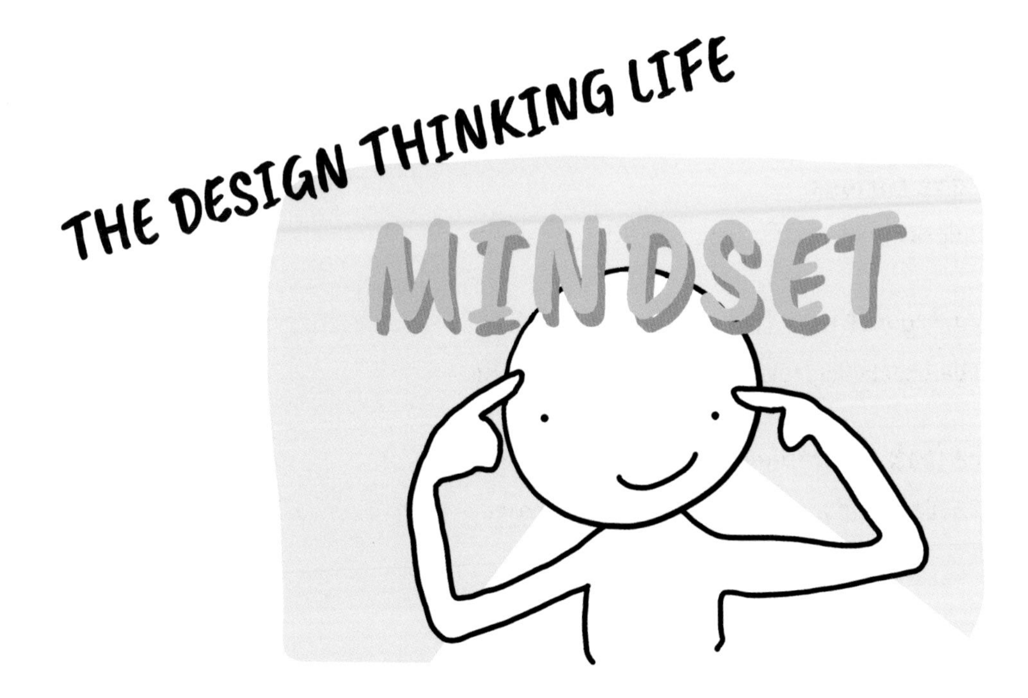

The design thinking mindset is characterized by curiosity, openness, collaboration, and by pragmatically trying things out.

The DESIGN THINKING LIFE mindset

We accept that we are on a journey.

It's less about the outcome than how we feel on the journey.

We let other people help us.

Personal development is a collaborative process because new ideas, insights, and feedback help us change our lives the way we want them to change.

We're curious.

Curiosity prepares our brain for learning new things.

We try out new things.

Experiments help test our assumptions and visions.

We look at problems from different angles.

Situations that are put in a new light make room for new solutions.

In the "DESIGN THINKING LIFE," we go through the phases: understanding, observing, defining point of view, ideation, prototyping, and testing.

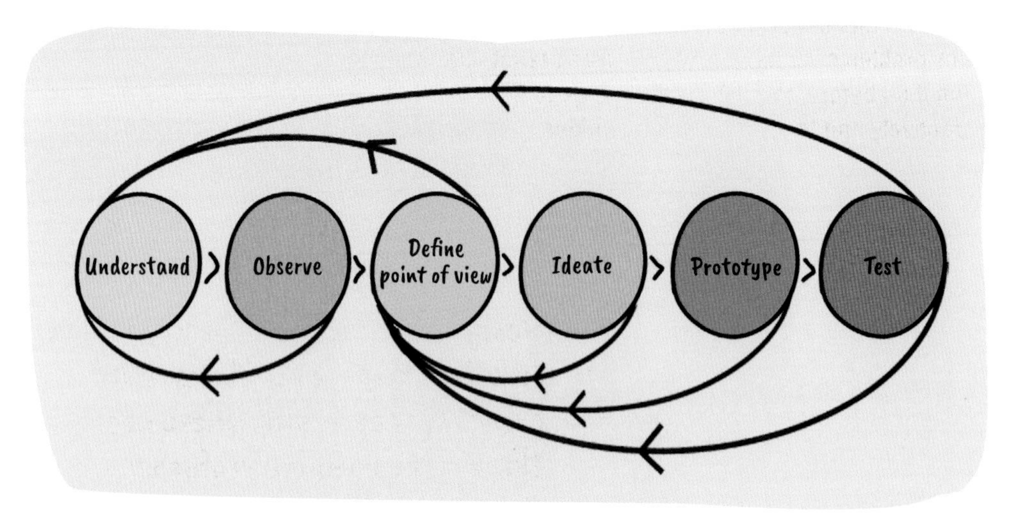

We will use this process again and again in the DTL book as guidance so that we always know where we stand.

In the beginning, the path to the desired change is unknown. The change takes place in many small, iterative steps.

A basic approach in design thinking is to place people with their needs and values at the center of our considerations. We have to fulfill tasks (jobs-to-be-done), undergo experiences that make us particularly happy (gains), and survive situations that frustrate us (pains). This focus on us humans is a core element; this is why design thinking is often referred to as human-centered design. So we try to get deep insights into people's needs. In the case of the "DESIGN THINKING LIFE," this means deep insights into ourselves. It is done predominantly with the help of self-reflection and others' perceptions of us that come from our social environment.

To deal with it now is actually optimal because at this moment a motivation for change already exists – not least because we deal with the topic of the "DESIGN THINKING LIFE." In other words: We can imagine living in a future that fulfills us and in which we feel better. We would like to illuminate from different sides problems we have recognized and react differently to them than before. We have the courage to think and act proactively and to seize our opportunities.

Design thinking has the aspiration to solve complex problems with creative ease – and where are there more complex problems than in our lives?

In taking stock, we assemble topics without evaluating them.

1) Which topics would you like to change, e.g., leisure, relationships, career?

We ask ourselves: Who am I? What do I like? What is working well, what not so well? What did I already try out in the past to change my situation? What has contributed to the fact that there has been some improvement?

In addition, we note down which skills and talents we possess, i.e., abilities that distinguish us from others. Maybe we are especially good listeners; or we enjoy working with numbers.

Top-flight athletes are especially good in a specific discipline, for instance. A shot-putter has the ability and strength to throw a heavy iron ball as far as possible. Her "circle of competence" is shot-putting. Hence, she will not win a marathon run.

Knowing what we can do well and where our gifts and talents lie is quite valuable, no matter whether we apply the "DESIGN THINKING LIFE" to our career, our health, or our relationships.

In this second reflection, we write down what we do especially well.

2) What can you do especially well?
Where do you have talent? What do you enjoy?

Although we cannot control life, we can have an influence on how we see ourselves!

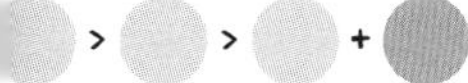

Top-flight athletes make use of another idea that is quite valuable for DTL. Besides self-reflection, they use a visualization of their athletic performance in their inner eye. For this reason, in the DTL book we will encourage you again and again

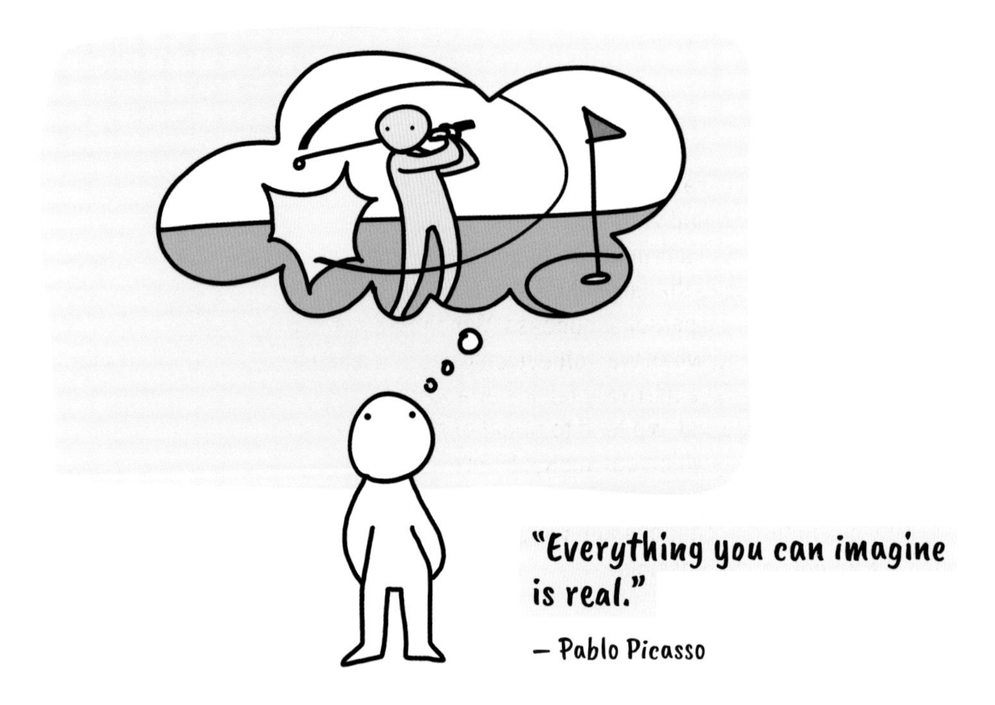

to visualize or sketch milestones and goals using your inner eye. This visualization helps us implement our wishes for change. At Stanford University, the phrase was coined:

"WHAT YOU FORESEE IS WHAT YOU GET."

In our example from professional sports, we would see ourselves on the top of the winners' podium.

The result of the "DESIGN THINKING LIFE" is an idea of what we aim at in the future and what helps to attain the desired goal.

At this point, we would like to show you a little trick so you will be really satisfied in the end with the results achieved. We know the feeling: We had a firm desire to achieve something, and at the end we missed the target by an inch, so we were dissatisfied – despite our personal success. We can frequently observe the phenomena when we achieve only second place in a sports event or when we narrowly fail in our application for team leadership. In general, we tend to orient ourselves upward and make comparisons, which can motivate us to become better. But often we feel frustrated because our current capability has natural limitations and we can influence it only within the bounds of the possible.

One suitable strategy for improving our motivation consists of orienting ourselves by looking downward in terms of our previous achievements and comparing ourselves to those who have achieved less. In our example from sports, second place may be our best result of the season. It's similar to a desire for a career change. For example: You may have already held your ground against other internal and external applicants and were very close to the target. You'll notice that you'll feel a lot better with this attitude. This mindset helps us especially when it comes to grades, ranks, best-seller lists, or a desired career goal.

After this brief introduction to the "DESIGN THINKING LIFE" and an initial stocktaking, we want to turn to our personas and subsequently design our own future. At the beginning of each exercise, Sue, John, and Steve will tell us something about their lives and their challenges, ideas, and solutions so that we can apply the tools without further assistance.

Who is Sue?

Seen from the outside, Suzanna – or Sue, as she affectionately calls herself – has a perfect life. She has a very good job with a bank in Hong Kong and, thanks to an assignment agreement as an expat, a very good salary to enjoy life in the former British colony. Nevertheless, she is unhappy with some aspects of her life. This is why she wants to change her life with the help of DTL tools.

Sue grew up in modest circumstances. Her parents didn't have the chance to attend high school, let alone a university. With a great deal of ambition and a strong will, Sue became the first member of her family to graduate from college. Almost thirty years ago, Sue's parents had come to Switzerland from Romania to work as seasonal help in the catering business. The family had a basic knowledge of the German language; the culture and the way of life in Switzerland, however, were quite different from the lifestyle in the small village in the Carpathian Mountains where Sue lived during the first years of her life.

Sue is very ambitious. She was a diligent student and graduated from high school as one of the top honor students of her class. After long discussions, her parents allowed their daughter to study at the university. Again, Sue graduated with excellent grades, although she had to finance her studies on her own with countless part-time jobs. After studying at the university, it was not so easy for Sue to enter into professional life. She applied for jobs everywhere, only to find out at the end of the process that other candidates had been chosen.

So Sue had no other choice than to complete another internship and, after that, accept a temporary position as a maternity leave replacement. Sue was able to build up a solid network in the company during the period of her employment and win over her bosses with good work, so she was offered a permanent position in the marketing department. After a few years in the job and completion of many internal training programs, her big opportunity came along. Sue was offered a management position in Hong Kong.

For Sue, it was a big career leap. Since then, she has had the privilege of experiencing five challenging years in Hong Kong and is increasingly aware of the fact that she misses Europe and her family. Moreover, Sue is still single. Although her open and direct manner allows her to meet lots of men, nothing has worked out thus far in terms of a longer relationship. Now, at the age of 35, she realizes that she wants to change her life. By means of self-reflection, she tries to figure out where she stands in life.

Self-reflection by Sue — where does she stand in life?

Through self-reflection, it becomes more and more clear in her mind that her desire for a partner, aspiration for a professional change, and longing to live again in Europe have top priority.

Now it's time to find out and define which points each of us would like to develop further: career, leisure, or our way of life – the possibilities are multifarious. In our introductory stocktaking on page 30, we already compiled topics. Now we'll evaluate them and put them in relation to one another in order to identify fields of action where our desire for change is strongest.

Possible fields of action may be relationships, leisure, work, and health. We will introduce more strategies and techniques starting on page 184 to deal with greater wishes for change in terms of professional life and career planning.

We start our journey with a pragmatic screening as to where we currently stand in life:

Self-reflection: Where do you stand in life?

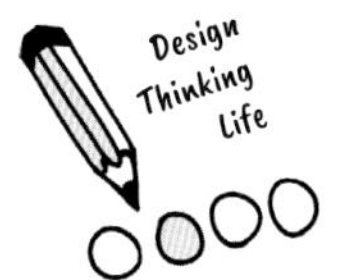

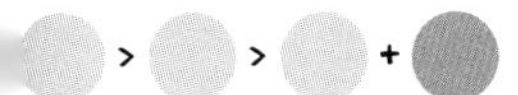

We can define additional categories or subcategories that are relevant to our life planning. Perhaps the relationship with our partner is undergoing a crisis just now because our personal wishes and the wishes of our partner are incompatible. In this case, we put the focus more on the topic of partnership and concentrate on specific subjects such as leisure, sexuality, family planning, communication, or fidelity.

According to what we know about Sue, she'll certainly tackle the issue of relationship. Next to the job situation, the fact that she has no permanent partner puts the greatest strain on her.

Do you have any specific things on your mind that you would like to change actively?

If you would like, you can put these subcategories in relation to one another again or use the matrix for other categories:

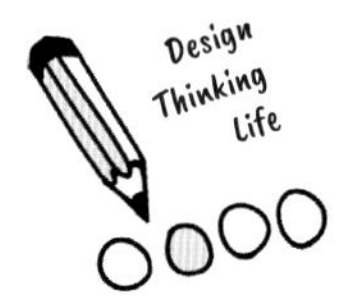

	😣	☹️	🙂	😍	😀
....					
....					
....					
....					

	😣	☹️	🙂	😍	😀
....					
....					
....					
....					

How everything is connected

From the previous phase of self-reflection, we can derive the themes and the points that we want to change. Sue, for example, would like a change in relationship and career. Usually, there are one or two sub-areas where we wish for a change. However, we can assume that all these areas of life are connected with one another. Imagine a mobile. Everything is held together in the system, and the individual elements mutually influence one another.

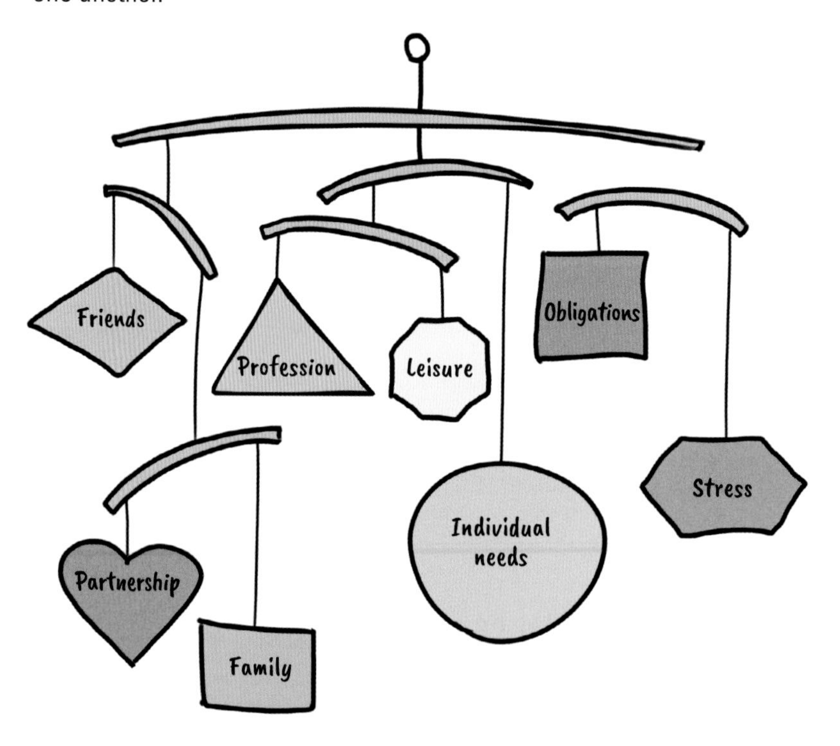

Dissatisfaction at work translates into physical ailments such as stress or headaches, which in turn might lead to insomnia. What's important here is that we first change our current life in small steps so that we still have the energy for greater changes. Often, a little thing kicks off something big or indicates the direction the change should take.

How do we find out what we want?

In design thinking, we work with empathy maps and in-depth interviews that help us learn more about a person. In the "DESIGN THINKING LIFE," we must go one step further and develop the ability to direct mindfulness toward ourselves and come to new insights by way of self-observation. The aim is to take a bird's-eye view (meta level). So we want to find out what we enjoy in life and which situations and experiences make us unhappy. This type of reflection helps us identify interference factors and unfavorable dynamics that may throw the system off balance.

For this reason, we expand the design thinking process defined in the beginning by two phases: acceptance and self-reflection. Now we have a total of eight phases that provide guidance and yield information on where exactly we are in the "DESIGN THINKING LIFE" cycle at the moment. We have integrated these eight phases as a color bar in the *DTL Playbook* so we always know where we are in the process.

We took stock and performed an initial phase of self-reflection at the beginning of the DTL book.

You asked the following questions: Where do I stand in life right now? What do I like? What do I want to change?

Self-reflection is a continuous process, so we apply this phase both at the end and at the beginning of the design cycle. This allows for a continuous development.

Reflect, accept, and understand

The first exercises have the aim of us learning more about ourselves. We find out which problems can be solved, who we are, what we do, and what experiences we have gathered. The findings we derive from these questions are unique; however, they help us understand our lives and ourselves better. In this context, reframing is an important tool because it helps us break out of existing patterns of thought.

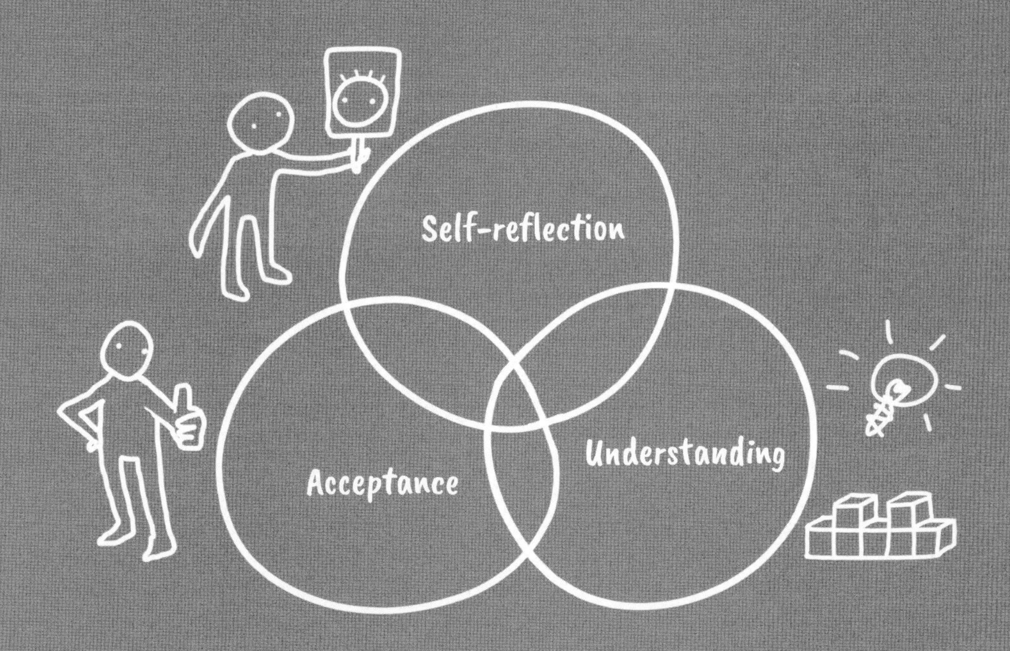

The difference between facts and solvable problems

The reason why we have integrated the "acceptance" phase in the DTL process and actually start with it is easy to understand. There are facts that we cannot change. Therefore, we must accept them or look at them from a different angle – i.e., perform so-called reframing – so that we can see these subjective problems differently.

Maybe it irritates us that we have to pay taxes, we think we pay too much for health insurance, or we don't like paying alimony. All these things can be annoying, but it's useless to expend energy on them because we cannot change them. Hence, we focus on solvable problems.

If it cannot be pragmatically changed, it is not a problem but a fact with which we have to come to grips.

For us to deal with these circumstances (the non-solvable problems) in a better way, we ask ourselves:

What has it got to do with me? Why does it stress me out that I have to pay such high taxes?

At the end of the day, we realize that it has nothing to do with us; taxes are simply part of the general framework of conditions. This makes us capable of acting again.

If you haven't made the distinction between facts and solvable problems until now, you should write a list so you can better classify the things that upset you personally. This way, the solvable problems can be quickly separated from the facts.

What upsets you?

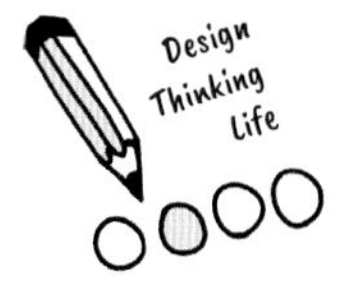

	Fact	Solvable problem

Responsibility means accepting that WE ourselves – every single one of us – are the cause and solution of our problems.

What is the meaning of reframing?

"Reframing" means nothing more than putting something in a new frame. We reinterpret something. The point is to see events, phenomena, or even information in a new context that is different from the one that first comes to mind. In design thinking, for example, we use reframing to reinterpret problems as opportunities/and market opportunities.

We all probably know the example of the glass of water. Is the glass half full or half empty? Both are objectively correct, but subjectively life is a lot nicer with the half-full glass. The half-empty glass, by contrast, is imprinted negatively on our subconscious, and negative thoughts are reinforced.

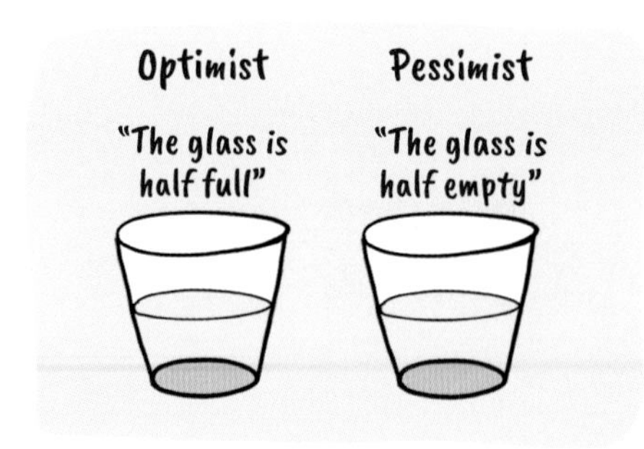

An optimist sees the bright side and thinks positively. The pessimist sees the dark side and thinks negatively. Both bear the consequences of their way of thinking – positive and negative.

Reframing also helps us deal better with the previously described facts that cannot be changed. Paying taxes, for example. We can fret over high taxes or can tell ourselves that we had a very good high school and university education and now we are earning good money and want to make a contribution so the system stays on a high level.

> **"It's not the things that distress us but the opinions we have of the things."**
>
> **- Epictetus**

At this point we want to solve a problem, also known as the 10-point challenge, before we reflect on how reframing can give us more vital energy.

Connect the following 10 points with a total of 4 straight lines without lifting the pencil.

(Answer on the next page)

Answer:

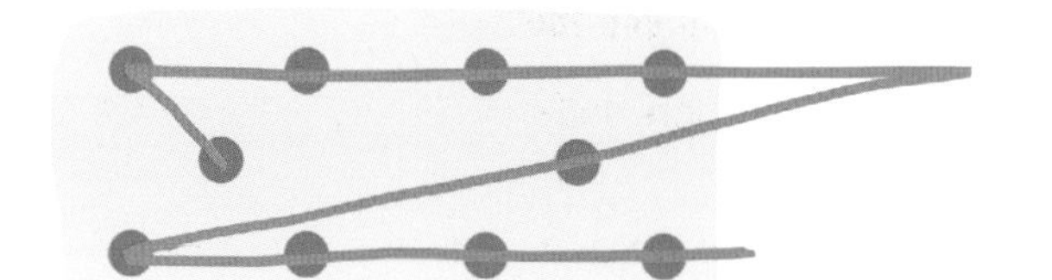

Break out of the frame to solve problems!

Reframing exercise

Is there a situation that makes you nervous? Sue, for instance, is afraid of giving a speech before large groups of people. Do you have a similar feeling in connection with a desired change or in relation to a fact? Think about how you might appreciate the feeling as positive for yourself, i.e., how to turn a threat into an opportunity. To do so, apply the three questions of this reframing exercise.

Situation: . . . .

. . . .

. . . .

1. Why do you have this feeling? What is its function? How can you reinterpret this feeling?

Sue, for instance, has the feeling of "stage fright" before a speech. As a result, Sue prepares herself especially well. During a speech, Sue's pulse rises. She gets red blotches, which indicate that her body is ready to give maximum performance.

. . . .

. . . .

2. How will you think about it in six months?

Sue is quite proud to have given such a professional speech. She has been asked twice since then to lecture on the same topic again.

. . . .

. . . .

3. What can you learn from the situation you feared?

Sue, for instance, learned that she got more self-confidence with each lecture she gave.

The path to self-efficacy

An important design paradigm for solvable problems is self-efficacy, i.e., the belief in our own abilities to change something instead of blaming coincidence, luck, or even other people.

The exercises on the following few pages aim at strengthening our self-efficacy. It is done by self-awareness, self-dialogue, self-reflection, and concrete action.

Start self-dialogue

Another tool for learning more about ourselves is taking an inner journey with the help of self-dialogue. To begin, we relax, close our eyes, and imagine a situation that made us especially happy. A one-of-a-kind tour or hike on a volcano, for example, where we had a marvelous view after the strenuous effort of climbing. But it can also be little things that make us happy. Enjoying a cup of cappuccino on the weekend, sitting on the balcony, browsing through the Sunday paper. Every moment that releases happiness hormones and triggers feelings of relaxation is welcome.

On the balcony, we sense the rays of the morning sun, a pleasant and warm sensation on our skin. Relaxing silence, only the birds are chirping, and it smells like spring. The chocolate powder on the cappuccino adds a sweet touch to the bitter taste of the coffee.

All situations that trigger emotions are stored better in our brain and can therefore be retrieved again faster.

Let's begin with our first self-dialogue right now. We close our eyes, relax, and imagine a happy situation. Subsequently, we write down what we felt as we were lost in these thoughts, feelings, or even physical sensations.

Notes from the positive experience

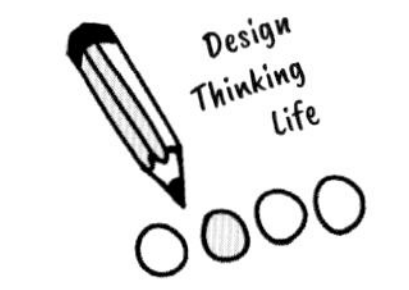

How did your body respond?
What did these thoughts trigger in you?
Where did you feel it?

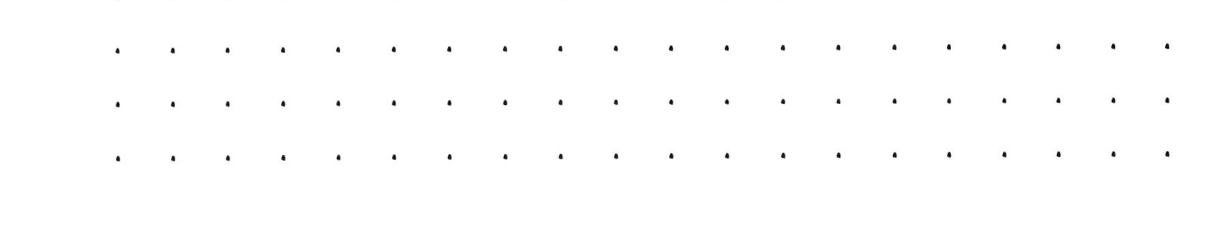

Now we do the same thought experiment but with an unpleasant situation. Afterward, we again write down what we experienced – feelings or even physical sensations.

This exercise can be uncomfortable, but we still recommend it because, besides the exploration of positive emotions, this type of self-reflection can sensitize your personal warning system. Situations that trigger negative emotions are a normal part of life; in the best-case scenario, we reflect on them so we can come to terms with them and finally accept them.

As in the previous exercise, we write down what we felt as we had these thoughts (feelings or physical sensations).

Notes from the negative experience

What did you feel?
How did your body respond?
Where did you feel it?

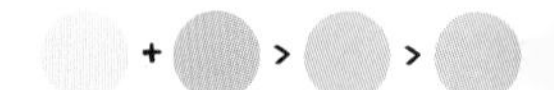

As already mentioned with the exercises we did previously, we can explore our personal warning system, as well as our positive emotions, and gain self-competence over our mind and our emotional experiential memory.

The emotional part sends its evaluation signals by means of images and feelings, which are consciously sensed by our body. This reaction happens four to five times faster than our ability to sort, understand, and evaluate something.

The mind evaluates experiences on this level: "Hey, that feels good – give me more of it" versus "That feels bad – scrap it."

Why in-depth knowledge of the warning system is important

The personal warning system helps us evoke positive associations in difficult situations and feel good with all our senses. Second, this exercise is a good preparation for the following concept that supports us in learning more about ourselves.

We've heard athletes say, "I was in the flow." For them, it is elementary to focus on something with utmost mindfulness and concentration. Mihaly Csikszentmihalyi describes "flow" as a state of total immersion. In other words: a feeling of positive intoxication with an activity that makes time fly.

Time flies

This state of total immersion in an activity in which we find an ideal balance between the challenge and our abilities can emerge when we're working, doing sports, making music, meditating, or writing a book. Flow activities usually yield a great deal of positive energy because we "exist in the moment."

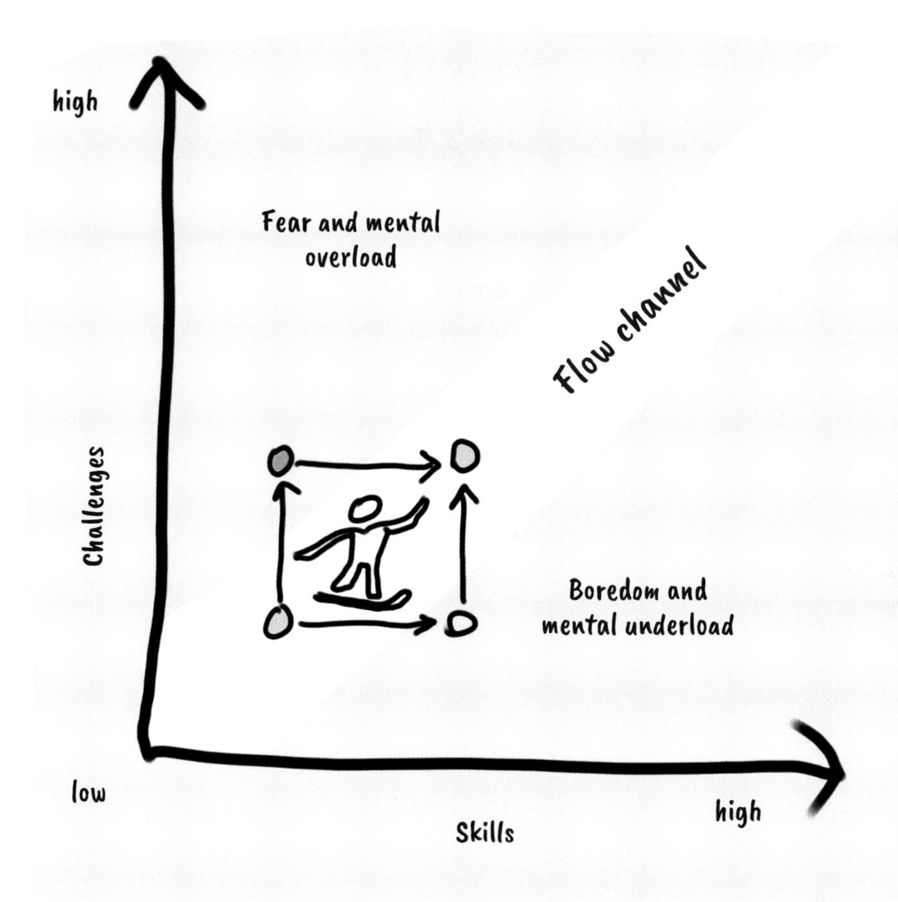

A good tool for finding out about our balance is keeping an energy journal over a period of one or two weeks. A two-week journal has proved to be of great value because it provides us with an opportunity to reflect on events and activities after the first week. This allows us to make little changes in the second week, to experiment with them and see whether a change we initiated makes a difference.

Keep an energy journal

What we would all like is to redesign our lives in just a few hours. Often, however, it makes sense to take more time and get to the bottom of our desire for change. Therefore, we recommend keeping an energy journal for the next one or two weeks. In the journal, we describe our daily activities, perceive the responses of our bodies, and kick off initial changes derived from them.

The energy journal

The notes in an energy journal help us observe ourselves and record what we especially enjoy and which situations or activities sap our strength or else give us strength.

On the following pages, the *DTL Playbook* provides you with an opportunity to document your states of mind over the next few weeks. We recommend you visualize your energy level (e.g., 0% up to 100%); write down the activity/situation (e.g., team meeting at work); document emotional signals (e.g., pressure in the chest area); and/or describe any emotions you felt (e.g., fear/frustration) and any responses of your body (e.g., stomach ache).

What is important here is constant reflection, i.e., the question: "How can I boost my energy level by 2% to 3%?" It's often the little things that can help here. A walk in the open air during your lunch break, for example, or a favorite song that you love to sing along with. It's important to know that relaxing and resting can give you a lot of energy because you refuel.

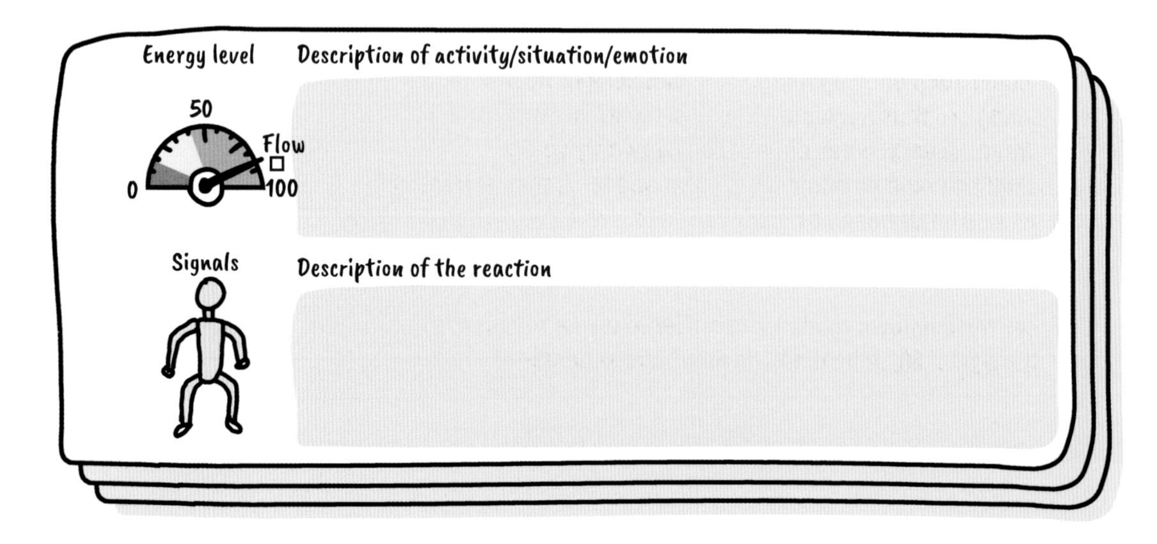

To understand working with the energy journal better, we would like to take a closer look at our fictional character John.

Who is John?

John, 60 years old

- Early retirement
- Married
- Children have moved out
- Lives in Memphis, Tennessee

Gains:

- Plenty of time for his hobbies

Pains:

- An unusual amount of time to spend with his wife
- High expectations on the part of his in-laws

Job-to-be-done:

- Make the best of the new phase of his life and enjoy retirement

John is just entering a new phase of his life. For the last twenty years, he's worked in the sales force of a mid-size company selling medical laboratory devices in the United States, from Florida way up to Maine. Due to his frequent traveling, John was seldom at home. Often, he came back from his customers late Friday night. His work week usually started on Sunday, so he could keep early appointments with customers on Monday morning. He thus neglected hobbies, family, his wife, and his children. Several times during the last twenty years, John wished he could be home more often and for longer periods of time. And he dreamed of riding his motorcycle more often, which was now parked in the garage for months on end.

When his employer made him an offer to take early retirement at the age of 60, John didn't hesitate and accepted.

Now that he is in early retirement, he's faced with new challenges. Sometimes he is overwhelmed by this new freedom; he's not used to being near his wife constantly; and he's dissatisfied with many activities during the week. To be more specific, there are activities now he had always been able to dodge in the past. His profession and the numerous trips as a sales rep were perfect excuses.

Now it's difficult to come up with good reasons for not wanting to visit the in-laws: John doesn't have appointments, nor does he have obligations. In the old days, John had sent his wife and children to visit, while he disappeared on one of his business trips. The current situation gets on his nerves. In addition, tensions between him and his wife have grown since he's been at home so much. They often argue, and for some time now John has felt that their relationship has become loveless and boring. Sometimes they don't speak a word to each other for hours on end because neither of them has anything new to say.

John would definitely like to redirect his relationship with his wife with the "DESIGN THINKING LIFE" and enrich his current phase in life with positive aspects. The last two days, John has made entries in his energy journal.

Entries in John's energy journal

Day: Sunday

Energy level

0 50 Flow 100

Description of activity/situation/emotion

- Motorcycle excursion with the local Harley Davidson Club
- Drive to Nashville via the beautiful Natchez Trace Parkway then lunch at the historic Casey Jones Village in Jackson

Signals

Some back pains in the evening

Joy

Freedom

Feeling young

Description of the reaction

- I felt free as a bird as we drove along the Natchez Trace Parkway with its many curves.
- During our break in Casey Jones Village, our motorcycles were admired by many passersby. My freshly cleaned Harley shone in the sun.
- At home, I arrived satisfied, although my butt was hurting, and I had a slight pain in the small of my back.

Day: Monday

Description of activity/situation/emotion

- Visit to the in-laws in New Albany.
- Accusations from the family that I have retired so early, with the children still being in job training and the house being loaded down with a mortgage.
- All that's important to me now, they say, is riding the motorcycle.

Signals

Feeling guilty
Frustration
Fear
Very heavy back pains

Description of the reaction

- I felt like a criminal. I felt guilty for having spent the entire Sunday with the Harley Club.
- Ultimately, I was feeling bad. During the argument, I sensed how tensed up I was. Then I had even more back pain.

John will continue to keep his energy journal during the next two weeks. For the next few days, he plans to meet up with former work colleagues, as well as introduce himself to a local blues and soul choir that he'd like to join. He noticed that relaxing moments such as morning rituals with coffee and music from the 1950s and 1960s give him energy as well.

Visualization in an activity/ energy chart

A very good and simple tool to analyze our notes from the energy journal is the visualization of the results in an activity/ energy chart. We would like to introduce it at this point before you start taking your notes to make it easier to see and understand what we are trying to achieve. The approach is as follows: After a week, you can arrange your activities on a timeline and evaluate them. This is done based on the energy level you have recorded in the journal.

John's example shows that his visit to his in-laws cost him a lot of energy (red bar) and that he probably experienced a flow state (green bar) when riding his motorcycle, as he smoothly sped around curves along the Natchez Trace Parkway and enjoyed the panoramic view.

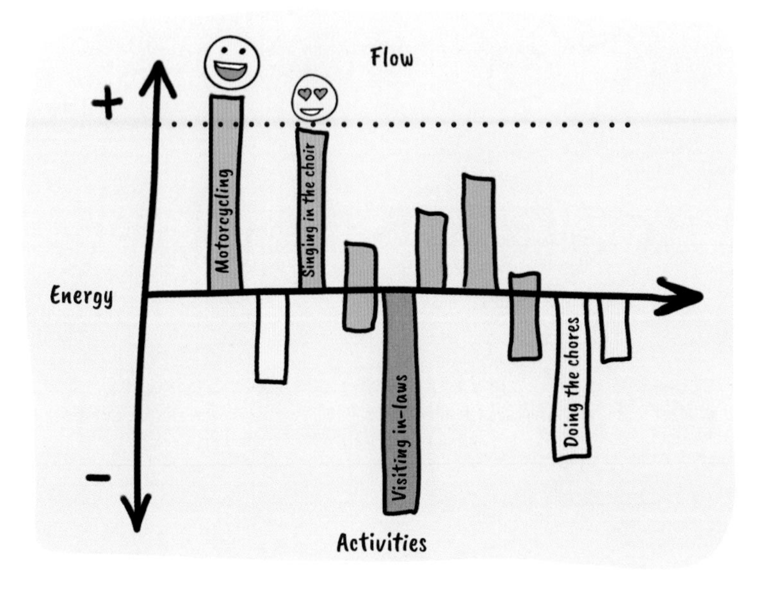

About 25% of the energy our body consumes is needed by our brain. It is therefore all the more important to use this energy for positive activities or to adapt the activities so that we perceive them as positive. We can also delegate activities we do not enjoy to third parties, such as annual tax returns or household chores.

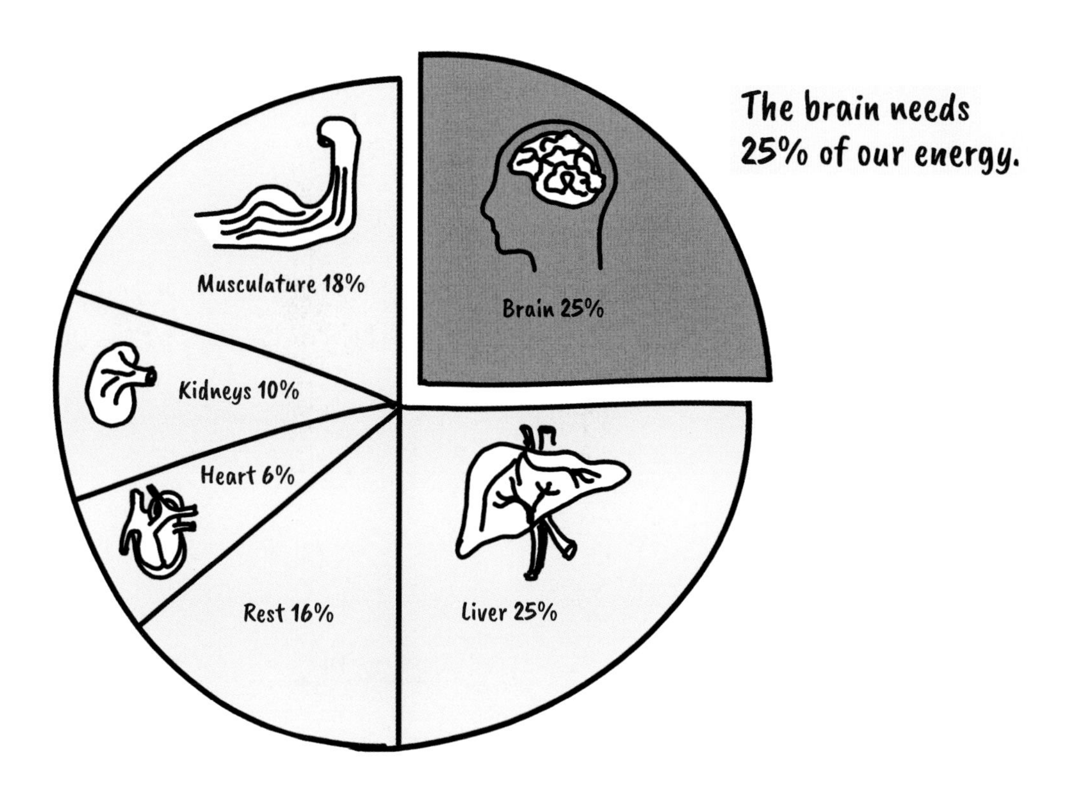

Before we begin the energy journal, we would like to briefly introduce another tool. The AEIOU grid, which is used in design thinking for observation, is excellent for identifying elements and situations that we can change. So it's a useful tool for reflection after you've kept the energy journal for a week, for example.

The AEIOU tool

AEIOU stands for **A**ctivities, **E**nvironment, **I**nteraction, **O**bjects, and **U**sers. This means, for example, that we can change the location or stay there for a shorter time than usual if we don't feel comfortable. In John's case, he might meet his in-laws by taking a trip with them, instead of visiting them at home. Or John might arrange for a cleaning lady to do household work from time to time so that this task is no longer necessary. He might also arrange activities he doesn't like in such a way that, before and after such chores, he does things that give him pleasure, e.g., riding his motorcycle to his in-laws (= reward).

AEIOU questions	
Activities	■ Which activities are fun? ■ What is your role?
Environment	■ Where do you feel comfortable? ■ How do you feel in a certain place?
Interaction	■ What interactions with others are enriching? ■ Who were you dealing with?
Objects	■ What did you enjoy? ■ What defined the experience?
Users	■ Who do you like to do things with? ■ Which people will help get you ahead?

Let's get started!

If you don't know what it is, then find it with the energy journal and make it a part of your life.

What happens in the next two weeks?

The following pages give you enough space to document and analyze your energy level during the next two weeks.
After one week, we recommend reflecting on the activities to create a first flow chart, to use the AEIOU as an aid, and to start the second week with positive changes. After these two weeks, you can analyze the previous week again, then reflect on and adapt the activities again, if necessary.

In general, there are two types of states that have a positive impact on our well-being. One is joy, ecstasy, and the flow activities already described. Second, there are states such as relaxation, sufficient sleep, gratitude, and serenity that help us replenish our energy tank.

Every day we need activities that we enjoy and make our heart sing.

Personal energy journal – Part 1

Your personal energy journal – first week

It is best to take some time every day to reflect on which activities and situations you have experienced. Visualize your energy level on the dashboard, enter your body signals, and describe the reaction. Use one, two, or more of the provided templates per day.

Date: __/__/____

Energy level

50 Flow

0 100

Description of activity/situation/emotion

Signals

Description of the reaction

Date: __/__/____

Energy level

50 Flow

0 100

Description of activity/situation/emotion

Signals

Description of the reaction

Date: _ _ / _ _ / _ _ _ _

Energy level

50

Flow

0

100

Description of activity/situation/emotion

Signals

Description of the reaction

Date: _ _ / _ _ / _ _ _ _

Energy level

50

Flow

0

100

Description of activity/situation/emotion

Signals

Description of the reaction

Date: _ _ / _ _ / _ _ _ _

Energy level

50

Flow

0

100

Description of activity/situation/emotion

Signals

Description of the reaction

Date: _ _ / _ _ / _ _ _ _

Energy level

50

Flow

0

100

Description of activity/situation/emotion

Signals

Description of the reaction

Date: _ _ / _ _ / _ _ _ _

Energy level

50

Flow

0

100

Description of activity/situation/emotion

Signals

Description of the reaction

Date: _ _ / _ _ / _ _ _ _

Energy level

50

Flow

0

100

Description of activity/situation/emotion

Signals

Description of the reaction

Date: _ _ / _ _ / _ _ _ _

Energy level

50

Flow

0

100

Description of activity/situation/emotion

Signals

Description of the reaction

Date: _ _ / _ _ / _ _ _ _

Energy level

50

Flow

0

100

Description of activity/situation/emotion

Signals

Description of the reaction

Date: _ _ / _ _ / _ _ _ _

Energy level

50

Flow

0

100

Description of activity/situation/emotion

Signals

Description of the reaction

Date: _ _ / _ _ / _ _ _ _

Energy level

50

Flow

0

100

Description of activity/situation/emotion

Signals

Description of the reaction

Date: _ _ / _ _ / _ _ _ _

Energy level

50

Flow

0

100

Description of activity/situation/emotion

Signals

Description of the reaction

Date: _ _ / _ _ / _ _ _ _

Energy level

50

Flow

0

100

Description of activity/situation/emotion

Signals

Description of the reaction

Date: __/__/____

Energy level

50

Flow

0

100

Description of activity/situation/emotion

Signals

Description of the reaction

Date: __/__/____

Energy level

50

Flow

0

100

Description of activity/situation/emotion

Signals

Description of the reaction

Date: __/__/____

Energy level

50

Flow

0

100

Description of activity/situation/emotion

Signals

Description of the reaction

First reflection with AEIOU

Make use of the AEIOU questions the first time you take stock of everything

AEIOU questions	
Activities	■ Which activities are fun? ■ What is your role?
Environment	■ Where do you feel comfortable? ■ What is the feeling like in a certain place?
Interaction	■ What interactions with others are enriching? ■ Who were you dealing with?
Objects	■ What did you enjoy? ■ What defined the experience?
Users	■ Who do you like to do things with? ■ Which people will help you get ahead?

Intermediate result – Where do you stand after one week?

Depending on the level of detail in your notes of the last seven days, you can evaluate your energy level for single days, a specific period (e.g., the weekend), or the entire week. We have inserted charts of various sizes so that enough space remains for the visualizations. For activities with an energy level over 95%, there is a high probability that they are flow activities, i.e., activities during which we forget time and in which we are completely immersed. But don't forget: Even brief moments of joy and relaxation can trigger a lot of positive things and release energy.

Visualization of the most important activities – When were you in the flow?

Time period: _______________

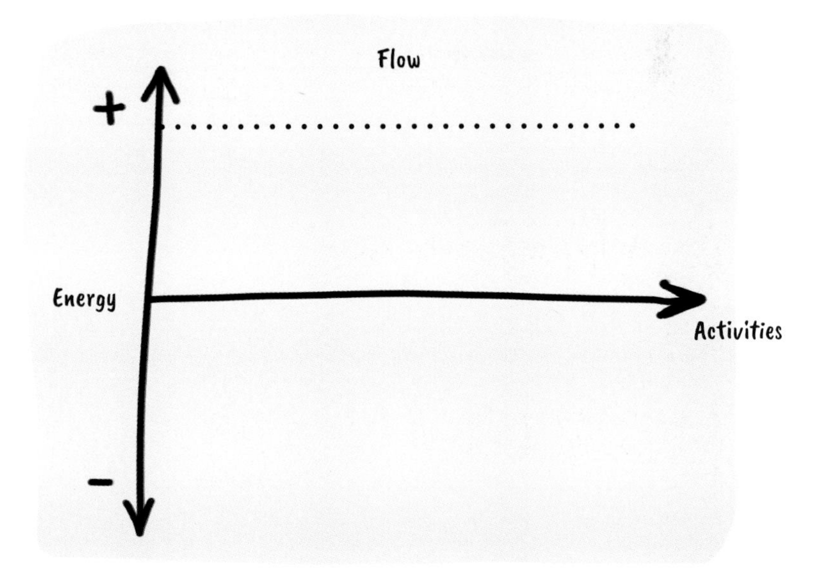

Intermediate result after 7 days

Time period: _______________

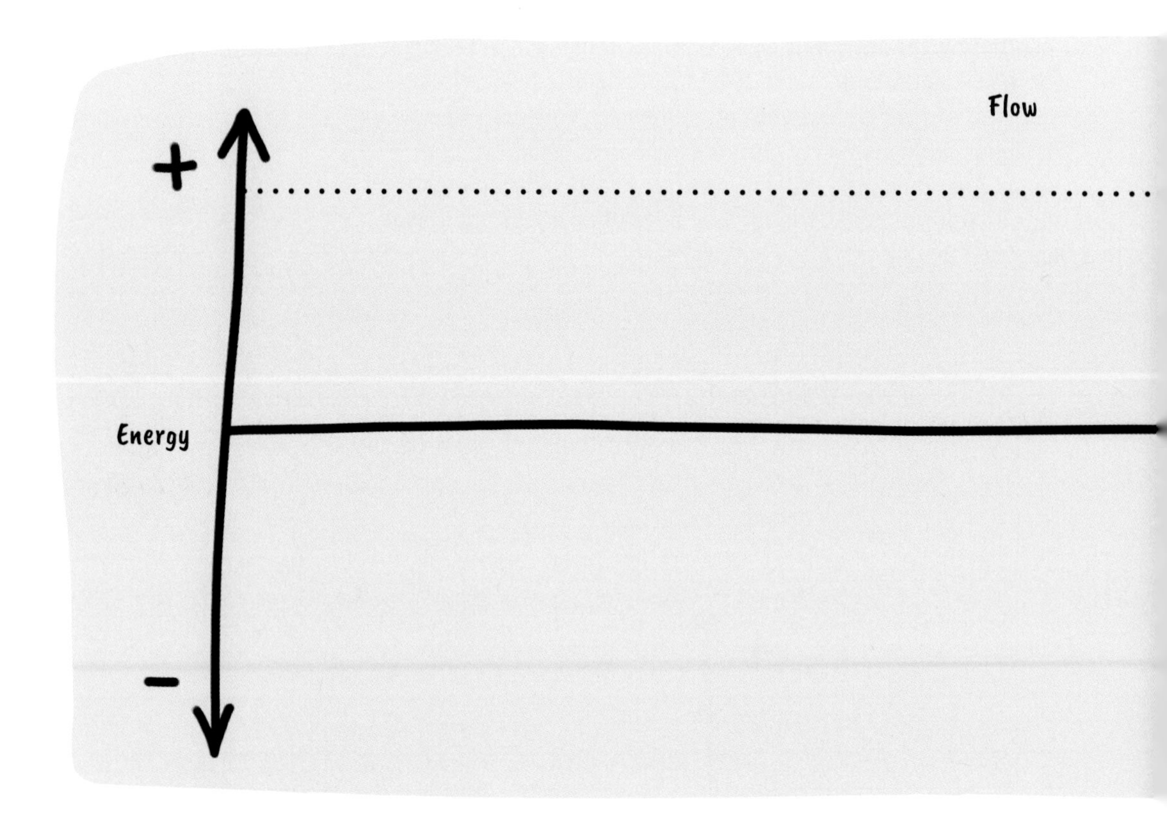

Activities

What do we focus on in the second week?

For the second week, you should expand small activities that previously gave you lots of energy and rearrange those activities that ate up lots of energy so that a first change can be initiated. Many things cannot be changed overnight; what you can do is have them take place at a different time or location (e.g., don't meet the in-laws at home) or reward yourself correspondingly with flow activities (e.g., go for a motorcycle ride before and after the activity).

What are the things you can change immediately?

Write down what you want to change, based on the findings from the energy journal.

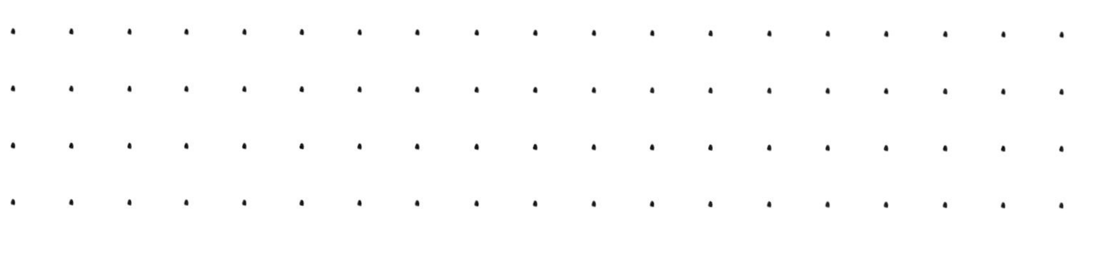

Personal energy journal – Part 2

Your personal energy journal – second week

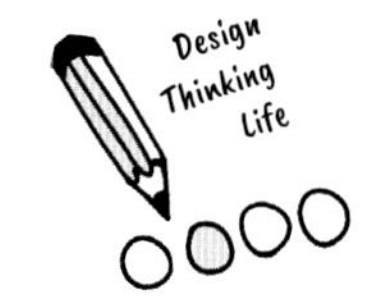

For the second part of your energy journal, you again take a little time each day to reflect on which activities and situations you experienced, including your energy level and body's signals. This way, you get initial insights into your changed well-being, e.g., when you have rearranged something in your daily routine.

Date: _ _ / _ _ / _ _ _ _

Energy level 50 Flow 0 100

Description of activity/situation/emotion

Signals

Description of the reaction

Date: _ _ / _ _ / _ _ _ _

Energy level 50 Flow 0 100

Description of activity/situation/emotion

Signals

Description of the reaction

Date: _ _ / _ _ / _ _ _ _

Energy level

50

Flow

0

100

Description of activity/situation/emotion

Signals

Description of the reaction

Date: _ _ / _ _ / _ _ _ _

Energy level

50

Flow

0

100

Description of activity/situation/emotion

Signals

Description of the reaction

Date: _ _ / _ _ / _ _ _ _

Energy level

50

Flow

0

100

Description of activity/situation/emotion

Signals

Description of the reaction

Date: _ _ / _ _ / _ _ _ _

Energy level

50

Flow

0

100

Description of activity/situation/emotion

Signals

Description of the reaction

Date: _ _ / _ _ / _ _ _ _

Energy level

50

Flow

0

100

Description of activity/situation/emotion

Signals

Description of the reaction

Date: _ _ / _ _ / _ _ _ _

Energy level

50

Flow

0

100

Description of activity/situation/emotion

Signals

Description of the reaction

Date: _ _ / _ _ / _ _ _ _

Energy level

50

Flow

0

100

Description of activity/situation/emotion

Signals

Description of the reaction

Date: _ _ / _ _ / _ _ _ _

Energy level

50

Flow

0

100

Description of activity/situation/emotion

Signals

Description of the reaction

Date: _ _ / _ _ / _ _ _ _

Energy level

50

Flow

0

100

Description of activity/situation/emotion

Signals

Description of the reaction

Date: _ _ / _ _ / _ _ _ _

Energy level

50

Flow

0

100

Description of activity/situation/emotion

Signals

Description of the reaction

Date: _ _ / _ _ / _ _ _ _

Energy level

50

Flow

0

100

Description of activity/situation/emotion

Signals

Description of the reaction

Date: _ _ / _ _ / _ _ _ _

Energy level

50

Flow

0

100

Description of activity/situation/emotion

Signals

Description of the reaction

Second reflection with AEIOU

Make use of the AEIOU questions to take further stock of the situation

AEIOU questions	
Activities	■ Which activities are fun? ■ What is your role?
Environment	■ Where do you feel comfortable? ■ What is the feeling like in a certain place?
Interaction	■ Which interactions with others are enriching? ■ Who were you dealing with?
Objects	■ What did you enjoy? ■ What defined the experience?
Users	■ Who do you like to do things with? ■ Which people will help you get ahead?

Intermediate result – Where are you after two weeks?

Depending on the level of detail in your notes of the last seven days, you can evaluate your energy level for single days, a specific period (e.g., the weekend), or the entire week. Again, we have inserted charts of various sizes so that enough space remains for the visualizations. And keep in mind: For activities with an energy level of over 95%, there is a high probability that they are flow activities, i.e., activities during which we forget time and in which we are completely immersed.

Visualization of the most important activities – Have you had any new experiences?

Design Thinking Life

Time period: ______________

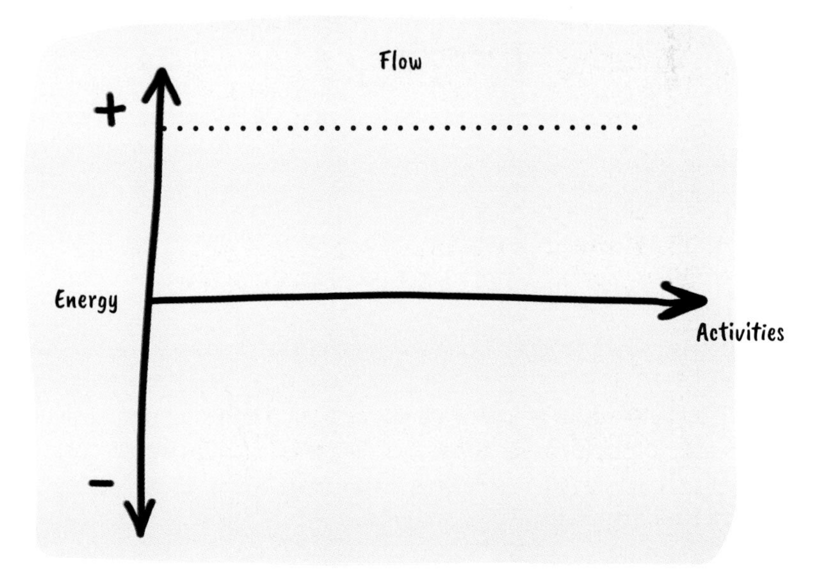

Status after two weeks

Time period: _______________

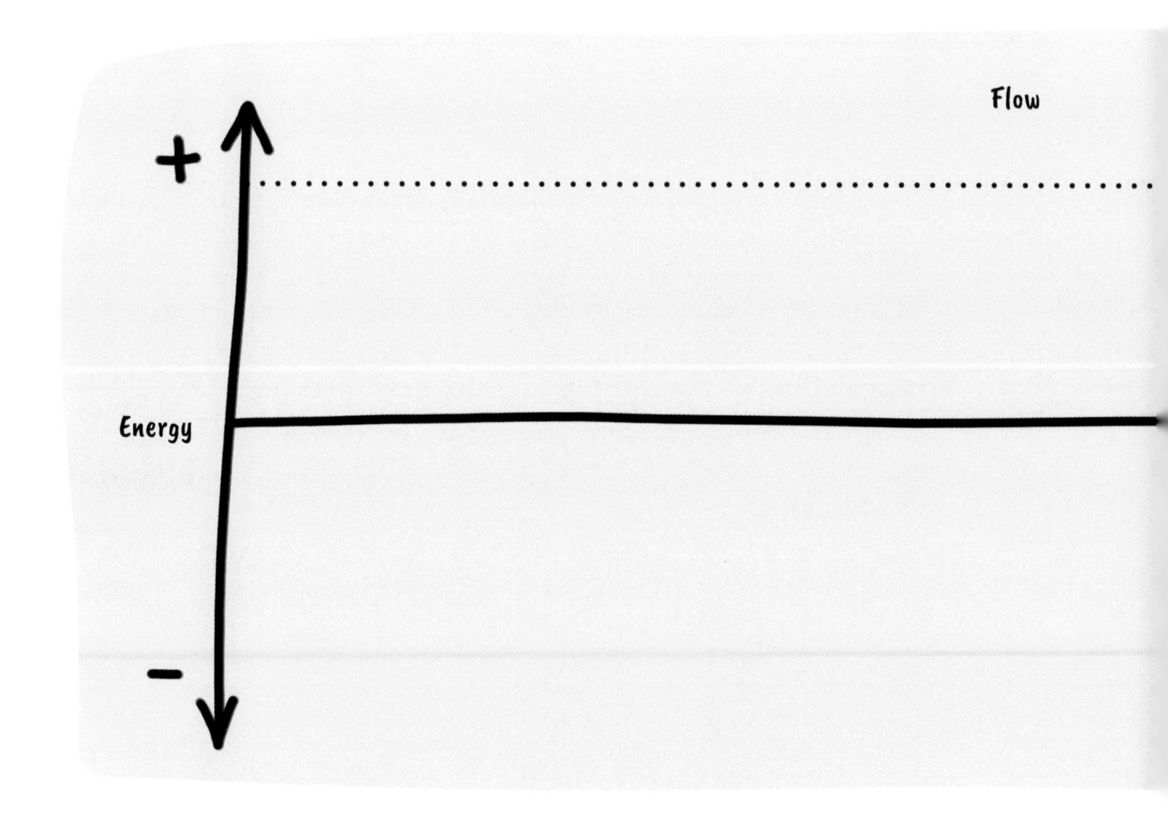

Are two weeks enough for an energy journal?

Depending on the phase of life you're in, you can keep the energy journal for a longer or shorter period of time. If you notice that your life is so complex that you need a longer period of time, we recommend you continue with the exercise for now and reflect on it weekly.

Activities

Cause-effect diagram based on the energy journal

If you want to examine your individual findings from the energy journal in greater depth, we recommend working with an Ishikawa diagram, also known as a cause-and-effect diagram. The impact you investigate can be both positive and negative. For example, John would explore what causes his insomnia and back pains in terms of his family and hobbies, but also which activities result in a flow condition.

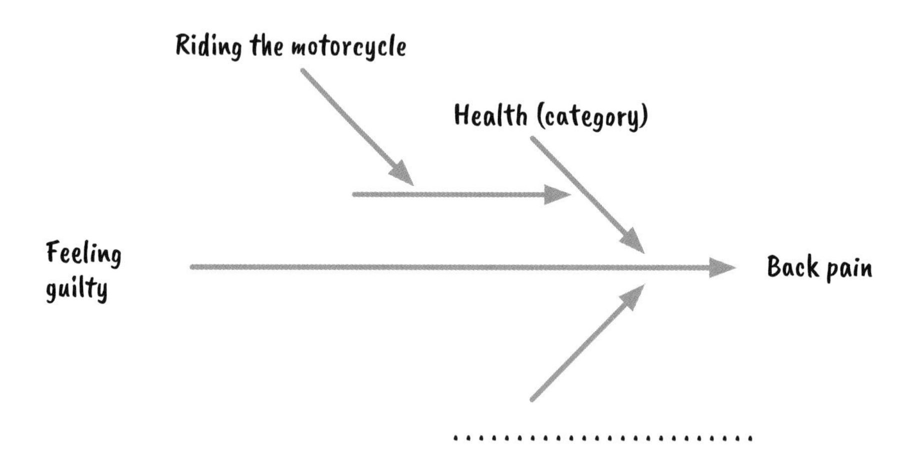

Design
Thinking
Life

Integrating other people's observations and perceptions of us

After we have learned more about ourselves with the help of self-reflection and the energy journal, the question now is what impression we make on others. The people we have contact with have a certain image of us. They perceive our actions and statements and thus form an idea of us. Others' perceptions of us are an important part of self-reflection because these perceptions provide us with information about our actual possibilities, strengths, and areas for development. In this way, we also obtain important suggestions regarding dynamics and starting points for change.

Where are we in the DTL process after the first few weeks?

Until now, you've dealt primarily with yourself. You discovered that it is important to accept facts, learn more about yourself, and observe how it feels if you alter sequences in your everyday life.

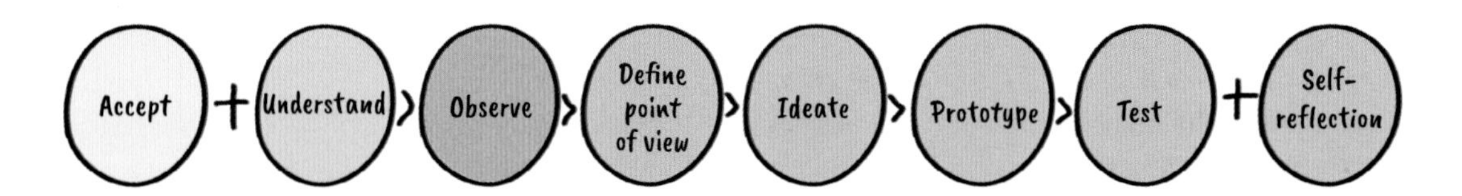

But since we live in a social system, we cannot ignore our environment; so it's important to include it in our considerations to better understand the impression we make.

For this reason, we'd like to go one step further and complement our self-image with others' perceptions of us in this section.

Interaction and communication are basic human needs that make it possible for us to illuminate the system and, at the end of the day, understand ourselves better. With the AEIOU questions, we've already begun to include the environment.

You have asked yourself which interactions are really enriching; whom you like do things with; and who the people are who help you get ahead.

These findings are important to complement your self-image with others' perceptions of you.

Do our actions give the impression that we are authentic?

Before we deal with the question of self-image and others' perceptions of us, we should pause and ask ourselves, "To what extent does our behavior give the impression that we are authentic or does it suggest that we only want to gain a better reputation in our social environment? In particular in social media, e.g., Twitter, Instagram, and Facebook, members of generations X, Y, and Z increasingly compete with one another through the number of likes, followers, and other rankings. These rankings are useful for self-marketing or product marketing, but they don't constitute valuable interactions with other people. We often turn ourselves into actors in a false scenario. The danger then is that we try to measure up to something we think we have to be like, even though we actually don't feel good in this role. The easiest way to become aware of this is to observe other people or groups: in selfies they take at parties, on vacation, beside the pool, and during dinner. It's not surprising that the images on Facebook always show the coolest party in town, the most attractive friends or the most beautiful places.

Unfortunately, these moments are often staged, and the actual interaction is superficial. We must acknowledge that real interactions with friends, a shared dinner, or relaxed hours beside the pool with your partner are of greater importance. So, maintain mindfulness with regard to the current interaction, and avoid compulsive worries about your reputation in social media, which quickly turn into a stress factor. In addition, the number of likes on social media says nothing about who we are or how we feel.

In order to dig a little deeper and learn more about ourselves, we will integrate our self-image and others' perceptions of us in the "DESIGN THINKING LIFE" reflection on the next few pages. As explained above, others' perceptions of us are about a perception of what we really are, not about a life that is play-acted.

There comes a point when the things themselves will not change but change will occur in the significance we give them.

Self-image versus others' perceptions of us

We have an image of ourselves, our subjective perception. In real life, it is of great value that we have a chance to ask others what kind of impression we make on them and what we trigger in them (empathy). In this way, we get feedback about ourselves.

One way to do this is to describe our self-image and ask ourselves how others perceive our behavior and what emotions it triggers in them. Then we can deliberately ask people in our environment what they think of us.

Our friends, colleagues, and family members can tell us what type of impression we made in a specific situation, how aspects of our personality are perceived, and which of our actions triggered certain emotions in them.

"The wise man learns from everything and everyone, the normal man from experience, and the stupid man knows everything better."

– Socrates

It is best to use this technique for a specific situation that we want to improve. In Sue's case, it might be the search for a life partner, a career change, or returning from Hong Kong to Europe. Sue approached the act of taking stock in two steps: First, she analyzed herself as a person – holistically, i.e., both her self-image and others' perceptions of her – and later she addressed the issue of a relationship (see the example on the following pages).

Self-image versus others' perceptions of us: Sue

Below, Sue's self-image is compared with her friends' perceptions of her. Sue marked the most significant differences and findings.

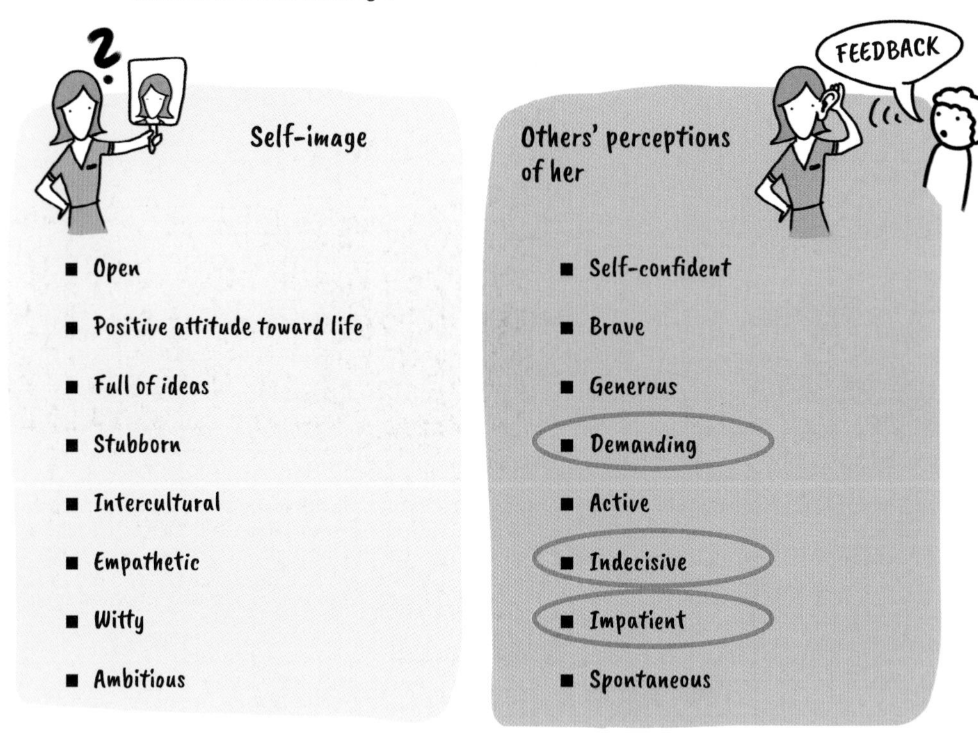

It usually doesn't take long for us to create our self-image. We have a clear opinion about ourselves. When we collect insights from others' perceptions of us, we need more time to ask friends, colleagues, family members, partners, and others in our environment to hold up a mirror to us. In many cases, we'll experience some eureka moments (highlighted in this example). These findings are quite valuable. Depending on the area of life, we should use WH questions to learn more. The question of who is suitable for giving us feedback depends on the change we're striving for (e.g., work, leisure, partnership, health). We play different roles in different areas of life.

Apply WH questions

Examples of WH questions: **When** was the last time I behaved that way? **How** did the people in my environment feel? **What** did you think when I behaved this way? Often, the exact answers to these questions offer possibilities for change. We should mark the most important among them at the end. The following pages provide space for this comparison and reflection. With “Self-image versus others’ perceptions of us,” we can compare the impression we make on various people or pick out specific topics.
Sue would like to take up the topic of relationship and partnership again, for instance, with good friends.

What is your view of the situation?
How do others see your situation, and what do they say about it?

Self-image

Self-image

Space for notes and new insights

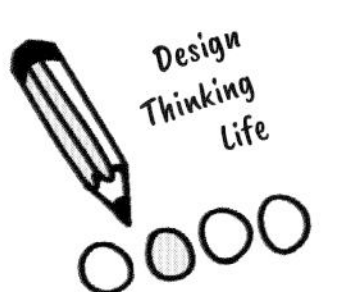

Define point of view

Through the various types of reflection, the energy journal, and the findings from "Self-image versus others' perceptions of us," we gained new insights about ourselves and got more familiar with our signal and warning system. Thus, we've already taken a major step forward in our efforts to change our lives. Based on these findings, we'll now define the point of view. This step is extremely important for the later phases in which we focus on the search for ideas for our change.

Define point of view

This means we leave the phase of understanding and observing and take another step forward in the "DESIGN THINKING LIFE" process. We formulate a point of view (PoV). In the context of "DESIGN THINKING LIFE," the PoV is expressed as a meaningful and implementable desire for change:

As the next step of change, I would like to

_______________________________________ (what?/mission)

because _________________ (need/reason/positive feelings).

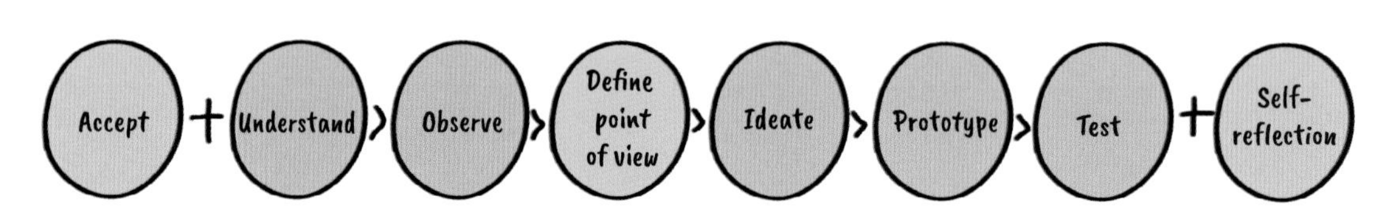

In this statement, we record what the most important findings are. First, we should sort out our thoughts and mark the most important findings from our self-dialogue, our activities, our self-image, and others' perceptions of us.

Create a context map

It always worked well for us to first present the findings in a context map that defines and visualizes the framework conditions. In this way, the thoughts on specific topics – e.g., relationships, work, leisure, and health – can be structured before we formulate the point of view.

Sue's context map on the topic of relationships

Sue has entered some findings on the topic of relationships ("Search for the right partner") into the context map. These findings come from the exercises on self-reflection (understanding and observing) she previously performed.

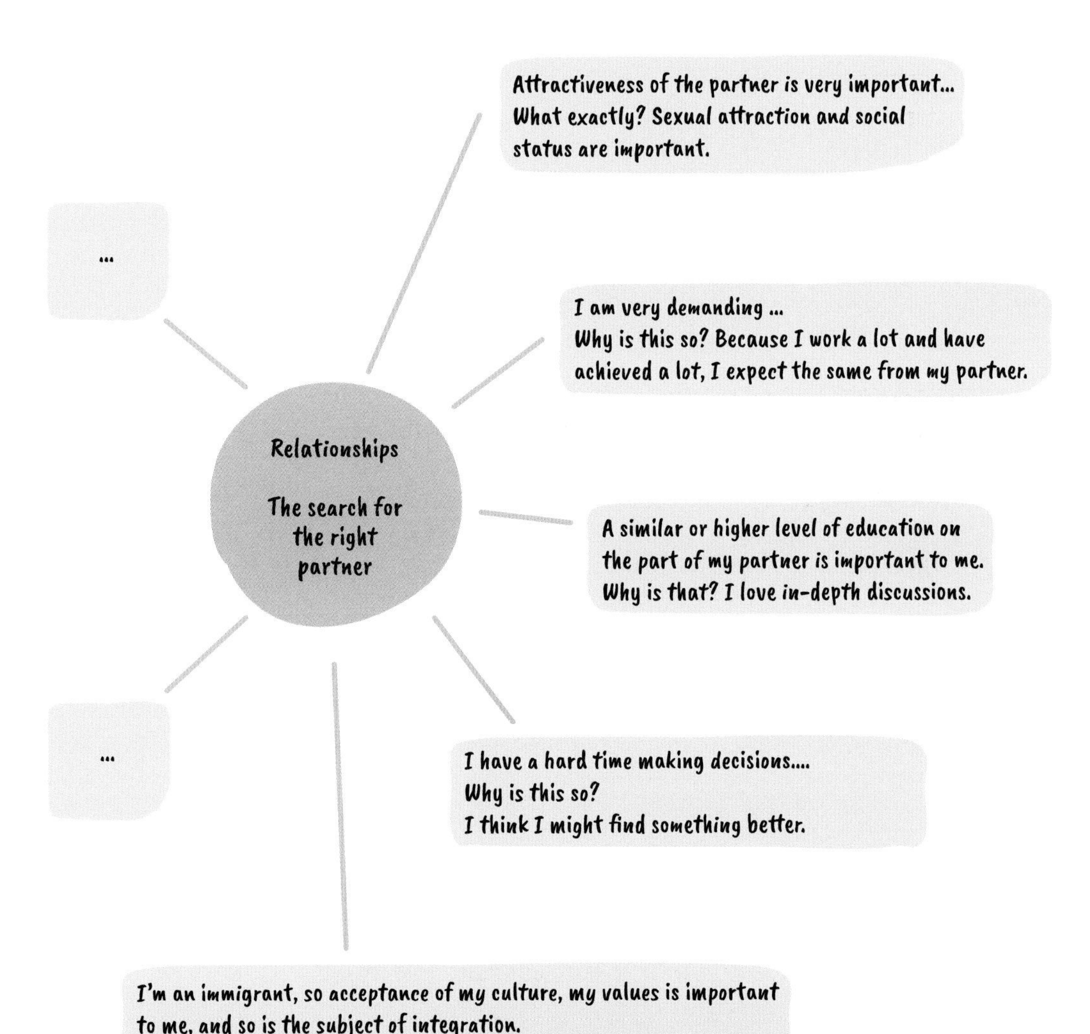

Create your own context map

There are several ways to create a context map. We may be able to add further ramifications. The important thing is that you put the crucial element in the focus of the considerations and write it down in the middle of the page. On the following pages, you have space to fill in your context maps before you formulate a point of view.

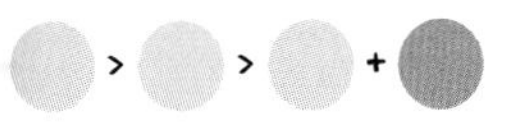

Design
Thinking
Life

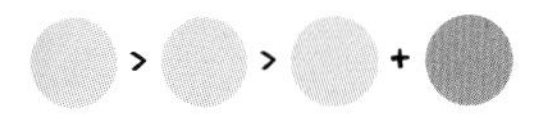

How do we formulate the point of view?

It's best to base the point of view on a visualization. In our imagination, we create an image of our desired future. We write this image down and describe it with any current sensations and feelings we experience. In addition, we imagine being happy in the context of our future change. This approach triggers additional sensations that we also write down.

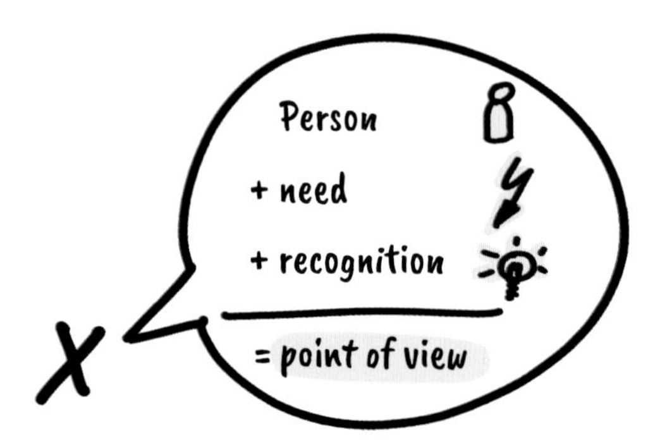

For example: In relation to her search for a partner, Sue has received feedback from various friends that she revealed too much of herself while getting to know each potential partner (a hypothesis) and thus scared men away. Up till then, Sue hadn't been aware of this. As an idea for change, her friends came up with a tip that she should try to appear a bit more mysterious.

Sue's point of view

Sue's point of view reads as follows:

"As the next step of change, I would like to appear more mysterious to men, so I don't scare them away while we are getting to know each other."

John was also able to discover something he would like to change through this step of reflection. John's friends speculated (hypothesis) that his wife might feel more loved again if he would express a little more joy in visiting his in-laws.

John's point of view

"As the next step of change, I want to do something together with my wife and my in-laws that we can all enjoy because it is important to me that my wife feels loved by me."

Now it's your turn to formulate – based on all your findings and the image of a potential future – a sentence that describes this state in the best way possible. We write the sentence in the present tense, in positive and simple terms. It is important simply to get started. We'll continue to edit our sentence in various iterations to arrive ultimately at the version we feel comfortable with.

Formulate point of view

When formulating the point of view in the DESIGN THINKING LIFE, we follow the formula John and Sue have applied. At the end, you should have a version you feel good with.

Version 1

"As the next step of change, ______________________________,
I (your name) would like to

______________________________,
(mission)

because ______________________________."
(need/reason/positive feelings)

When you look at your sentence again now, you may reflect on it:

- Are you really satisfied with the statement?
- How does it feel when you're on the move with this idea?
- What exactly feels better? Where do you feel it in your body?
- How can others recognize your new mission?
- What would you do more of? What are the interferencial factors that bring about a positive feeling?

Try to find the right place for the sentence on the following scale.
If you enter it more toward the left, you should adapt it and improve it until you have a good feeling about it.

We have inserted the formula several times to leave space for iterations. Often, we get it right immediately with the first or second version; sometimes only a little adaptation is needed to describe a point of view that we feel good with.

Improve point of view iteratively

Version 2

"As the next step of change, ______________________________,
I (your name) would like to

______________________________,
(mission)

because ______________________________."
(need/reason/positive feelings)

Version 3

"As the next step of change, ______________________________,
I (your name) would like to

______________________________,
(mission)

because ______________________________."
(need/reason/positive feelings)

Final point of view

"As the next step of change, ______________________________,
I (your name) would like to

______________________________,
(mission)

because ______________________________."
(need/reason/positive feelings)

Find and select ideas

In DTL, we want to develop a great many ideas and possibilities to demonstrate the whole range of ways to become self-efficacious; only then will we select individual ideas for testing. In this *DESIGN THINKING LIFE Playbook*, we use creativity techniques, such as brainwriting and the use of analogies, to get new ideas. These are effective ways to explore new life paths.

How do we ideate (find ideas)?

After the understand and observe phase and formulating a point of view, it's time to invest our mission with definite ideas. The starting point for this is our point of view from page 108/109. We begin with an active search for ideas and initial approaches to a solution.

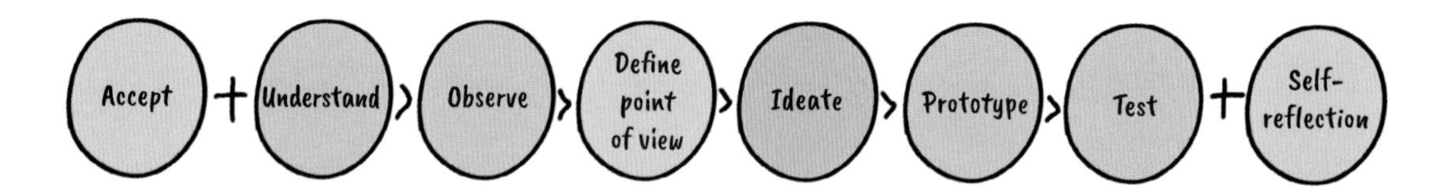

Perhaps during the last few days and weeks, initial ideas have already occurred after our various reflection exercises on how we can initiate and implement our change.

We may have already implemented or experimented with some of the changes in the first few days as part of our reflection on the energy journal. But since we shouldn't change everything all at once, we now focus on the elements that we think are particularly important or that we feel are urgent.

We know the reason for this. Our life can be seen as a mobile in which individual areas balance one another out, and each change leads to finding a new balance being found in the overall system.

In particular, we should call to mind what are unchangeable facts and what are problems we can tackle.

The criteria to select the change must be defined by each individual on his or her own. In many cases, it can also be a path that has different stages, similar to climbing a high mountain. Small stages can be derived from the energy journal, for example. Don't worry if things slow down at a certain point. In this process, taking small breaks (standstill) to pause for a moment and reflect are quite normal and very valuable. A rest helps us gather strength for the next stage.

Each small stage that we reach strengthens our self-confidence and self-assurance, so that, over time, we will be able to tackle new peaks and challenging passages. We grow with the challenges we face.

Sue's stages, in which she searches for ideas and solutions

Sue has formulated "Appearing mysterious" as the first stage.
A second stage might be to date men of a different social status.
We are focusing on the "WHAT" in formulating the stages. The "HOW" would be describing the solution.

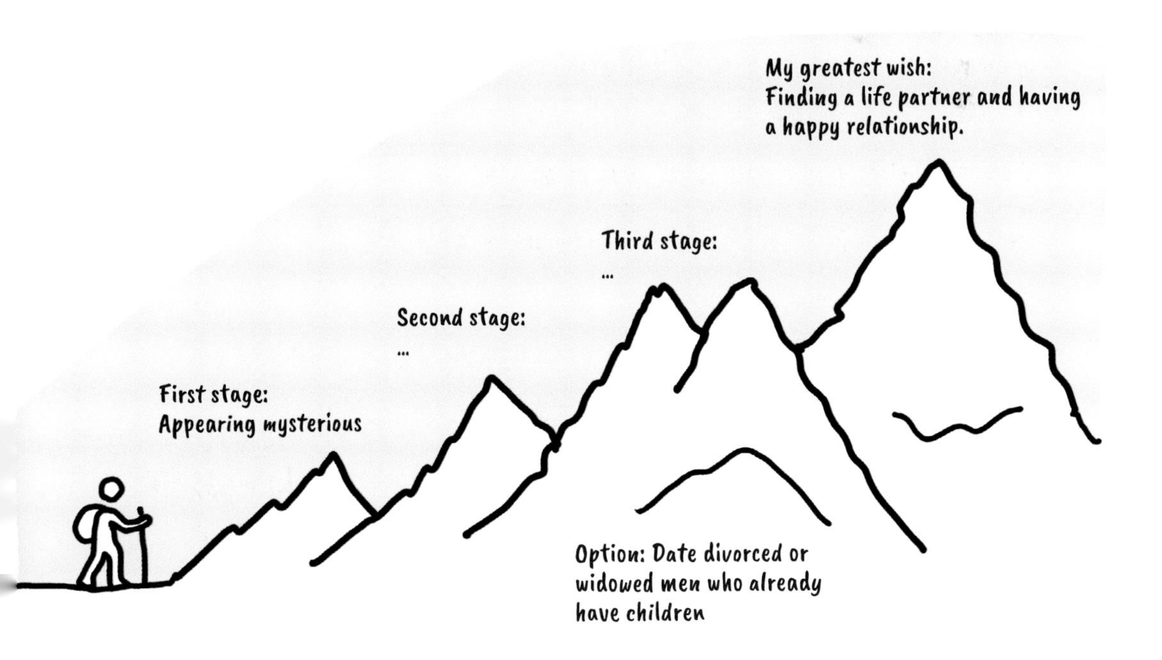

Define stages and goals

Define your stages and goals for one or more areas of life.

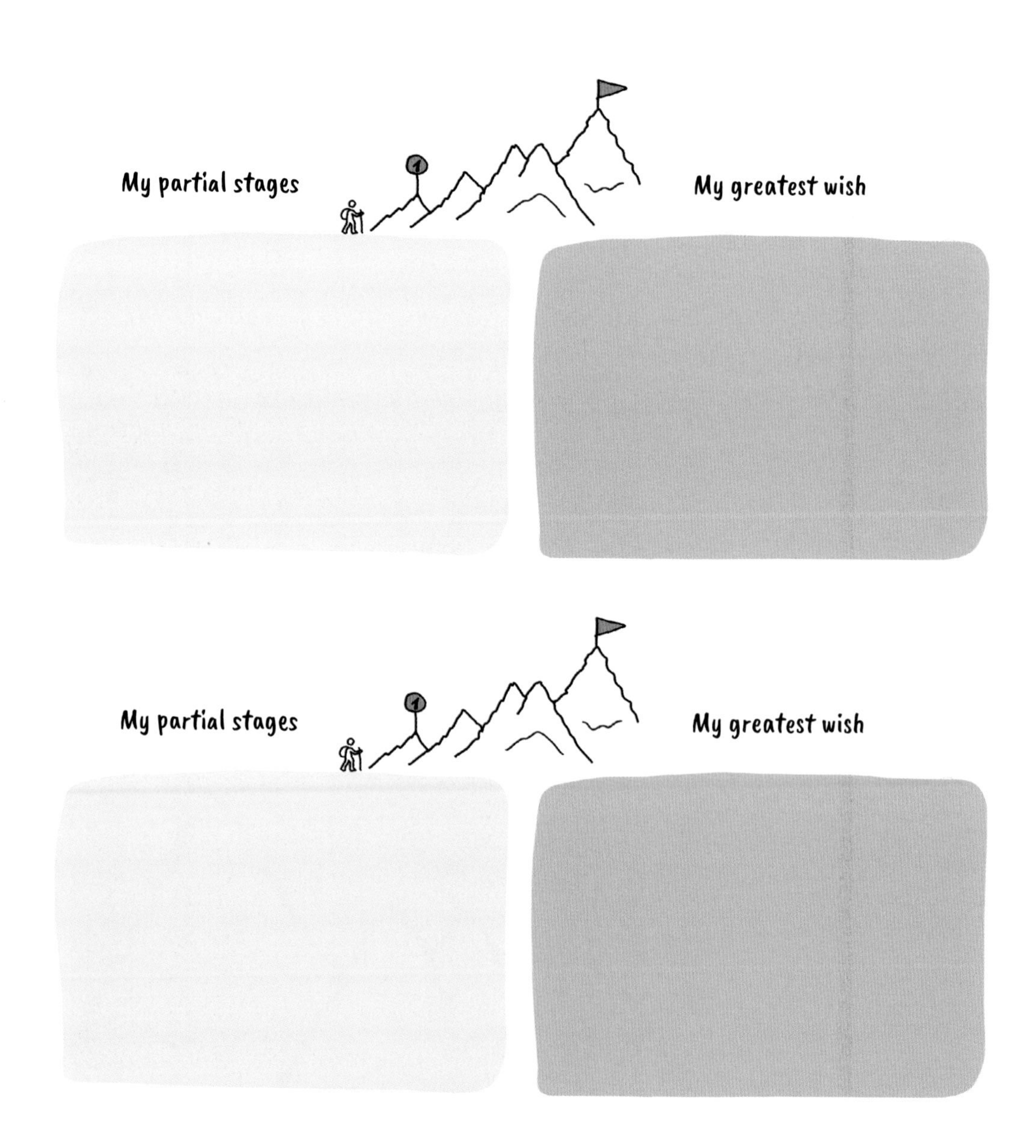

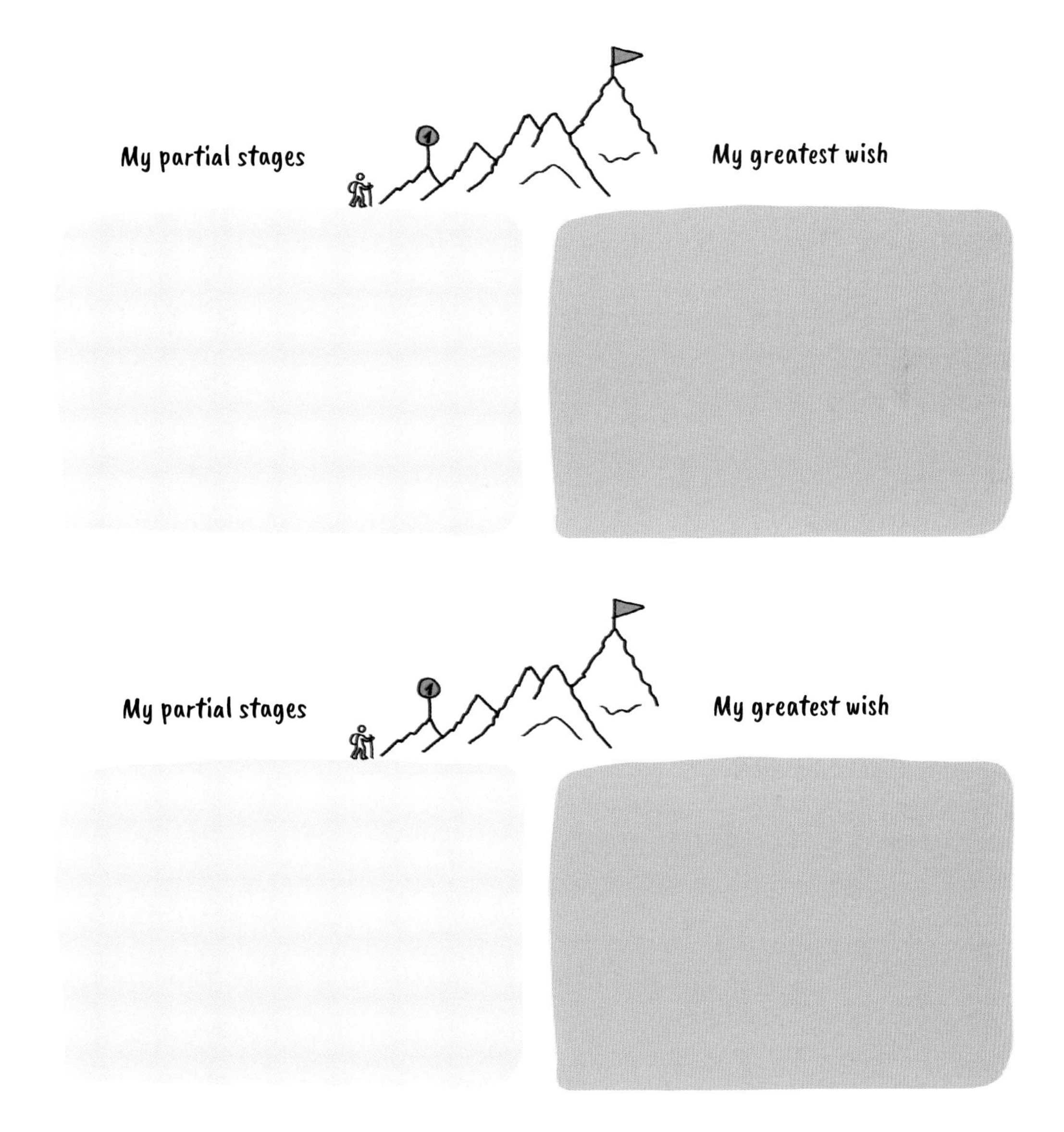

After we know "WHAT" we want to change, we are in need of ideas for "HOW" to make this change.

How can we develop creative ideas to realize the first stages?

There are two states in design thinking: diverging and converging. Diverging is about generating as many ideas as possible. In order to ideate (find ideas), we use a well-known creativity method; namely, brainstorming or brainwriting (writing down ideas yourself). Converging, by contrast, is used to focus and sharpen the most suitable approach to a solution by trying out new concepts and finally implementing them.

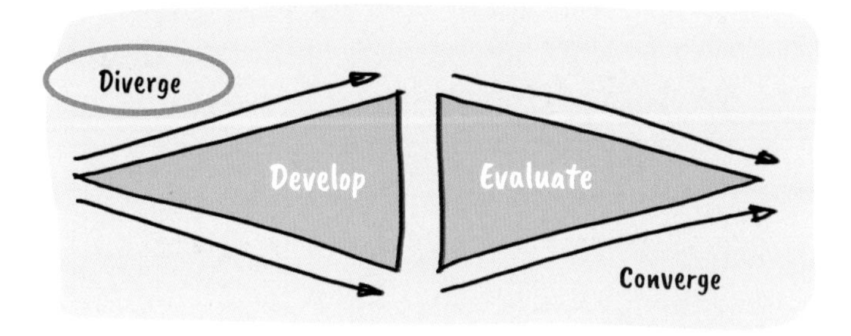

In this phase, we want to give free rein to all ideas. Brainwriting is ideal for getting started. We write down everything that comes to mind.

John's first ideas for the stage "Give the relationship more momentum."

We remember John. In the first stage, he wants to improve his relationship with his wife. He feels the relationship is boring and loveless. Their common desire is to have a happy relationship. He starts with a brainwriting session and notes all the creative ideas that come to his mind regarding this stage.

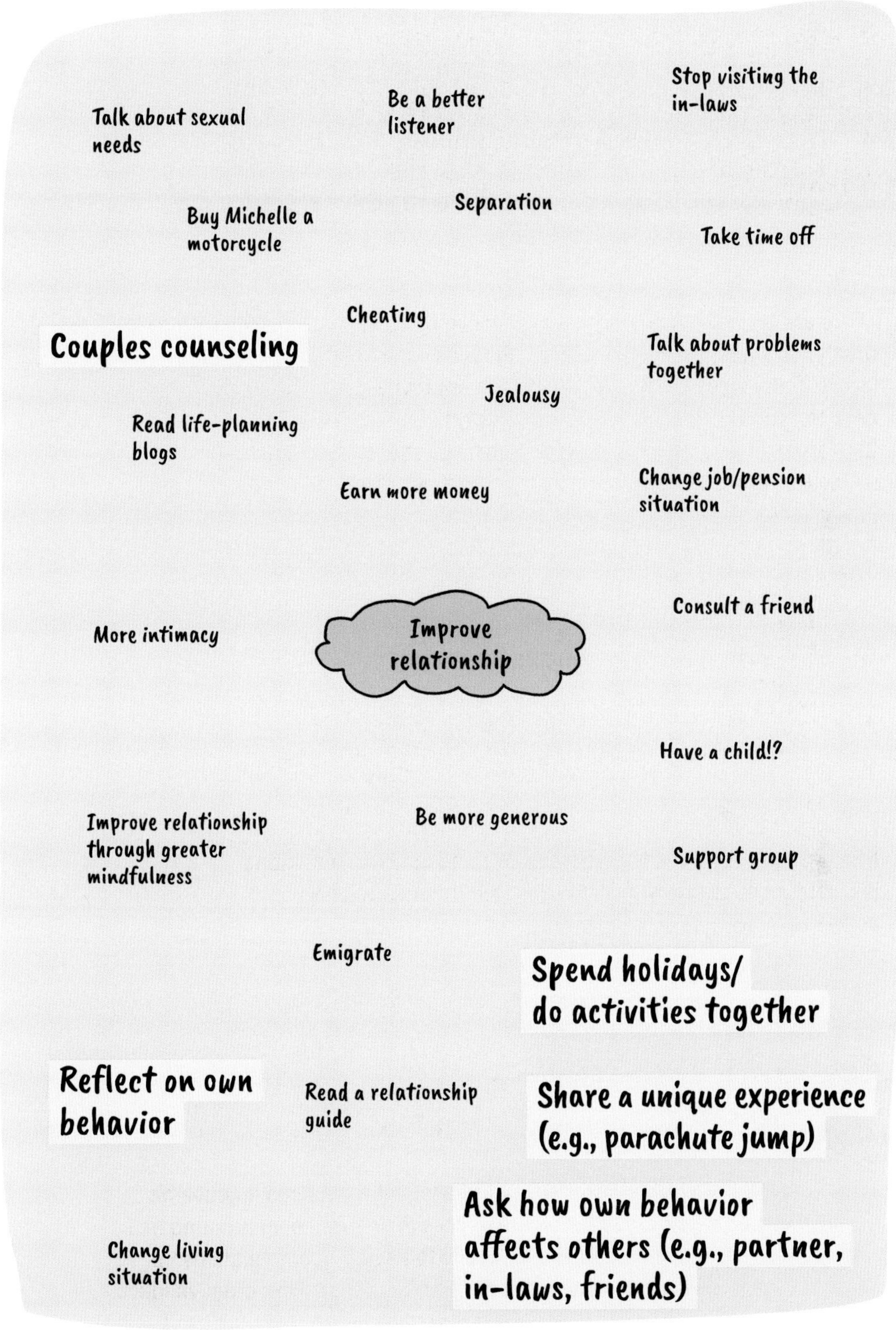

Talk about sexual needs

Be a better listener

Stop visiting the in-laws

Buy Michelle a motorcycle

Separation

Take time off

Cheating

Couples counseling

Talk about problems together

Jealousy

Read life-planning blogs

Earn more money

Change job/pension situation

Consult a friend

More intimacy

Improve relationship

Have a child!?

Improve relationship through greater mindfulness

Be more generous

Support group

Emigrate

Spend holidays/ do activities together

Reflect on own behavior

Read a relationship guide

Share a unique experience (e.g., parachute jump)

Ask how own behavior affects others (e.g., partner, in-laws, friends)

Change living situation

Sue's list of ideas for finding a life partner

Sue has also ruminated on **"HOW might I..."** . She is looking for ideas to find the right life partner. In a first step, she wants to figure out the occasions when she can test the insight of "appearing mysterious." Sue uses a table for brainwriting.

Brainwriting from Sue about "On what occasions can I appear mysterious?"	
■ Get to know someone at a new leisure activity	■ Sign up for speed dating and get many contacts in a short time this way
■ Take a class at the adult education center	■ Write poems and publish them online (e.g., as a blog) or read them in public
■ Rehearse and enact a kind of "street performance" to appear mysterious	■ Participate in events with a mixed and international audience
■ Study something else with no relation to economics and marketing	■ Go on a study trip for singles
■ Accept invitations from friends more often and attend more private parties	■ Go to meet-ups on topics you don't know anything about
■ ...	■ ...
■ ...	■ ...

Ideate

We also want to brainstorm creative ideas, so we use the following pages to write in our achievable wish or a first realistic stage with ideas. Let's do it as John did. We write the concept of the stage in the middle of an empty page in the DTL book. Then we write down all the associations (wild ideas, ideas that are absurd at first glance, radical ideas) or work on a list with two columns, as Sue did.

If things slow down, we simply continue elsewhere. It's about the amount of ideas, not about quality, and if we don't get ahead, we ask ourselves: "**HOW might I ...**"

Brainwriting: What are your initial ideas for how to realize your stage?

Brainwriting and brainstorming for your ideas

Design
Thinking
Life

Brainwriting and brainstorming for your ideas

Design
Thinking
Life

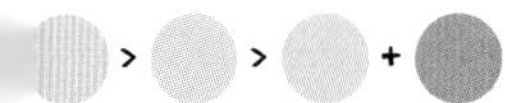

Select ideas

No doubt you'll have a lot of different ideas for one or more stages after this exercise. Highlight five to seven ideas with a highlighter. You should intuitively select the ideas that you find helpful and supportive and transfer them to a "daisy." It is important to make this initial selection because it will make you self-efficacious and trigger a positive change.

The advantage of working with a daisy as a visualization method is that all the petals of the flower are equal, and you will not automatically prioritize ideas, as is often the case when you use lists.

Tamara Carleton and Bill Cockayne from Stanford University, for example, deliberately use the daisy in their Foresight Framework to show the possible dimensions of a problem or, as in our context, to map a certain number of possibilities.

Sue's selection of ideas

Sue has come up with more than twenty ideas through brainwriting on how she can "appear mysterious." Some have hit the mark with her, others less so. For her goal of "appearing mysterious," she selected the following six activities.

Selection of ideas for your stage

On the following pages, there is space to create your own daisies and sort your thoughts. It is important to select ideas that you feel good about and that you believe are useful for reaching the stage.

Design
Thinking
Life

Design
Thinking
Life

Are we satisfied with the ideas or are we looking for even more creative ideas?

After this first iteration of ideation, it may be that we are already quite satisfied with the result. Then we should go straight to page 138 and initiate the transition from a diverging phase to a converging phase.

If we want to boost our creativity and ideate more (generate more ideas), we should try another design thinking method; namely, working with analogies. The basic idea here is to incorporate helpful ideas from other areas or phases of one's life into the brainstorming process in order to come up with new solution variants.

What analogy might Sue choose?

Sue's first thought with regard to "appearing mysterious" is the analogy "secret agent." For her, James Bond is the epitome of someone who appears mysterious. Even the well-known statement "Shaken, not stirred" awakens interest and makes the MI6 agent intriguing.

Sue formulates the following sentence from the chosen analogy:

How can I improve my current behavior (which is too assertive and transparent) by behaving like a secret agent?

John gives us another example. Analogies from the past can also be used. John remembers the time when he met his wife, Michelle, and he develops associations on this basis.

Analogies for John's stage

"Why Michelle and I were so much happier when we said yes thirty years ago."

OR

"What were the qualities I so appreciated about Michelle when we met?"

Brainstorm and describe analogies

When working with analogies, it is useful to think up two or three analogies and then formulate them as a sentence. After that, we recommend you carry out a brainstorming session for one or two analogies.

Brainstorming for analogies:

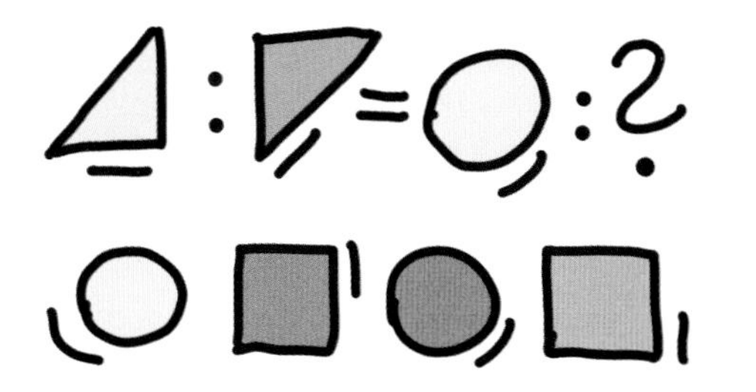

Analogies help us change the perspective and boost creativity.

Analogy 1

How can I, by way of including ______________________ (benchmark),

improve my current situation ______________________ (problem)?

Analogy 2

How can I, by way of including ______________________ (benchmark),

improve my current situation ______________________ (problem)?

Analogy 3

How can I, by way of including ______________________ (benchmark),

improve my current situation ______________________ (problem)?

How do we ideate based on an analogy?

In the context of brainwriting, we write down all possible solutions for the chosen analogies on the following pages and then enter the most suitable solutions into our daisy again.

After her brainstorming session, Sue has already identified various solutions with respect to the James Bond analogy and summarized them in another daisy.

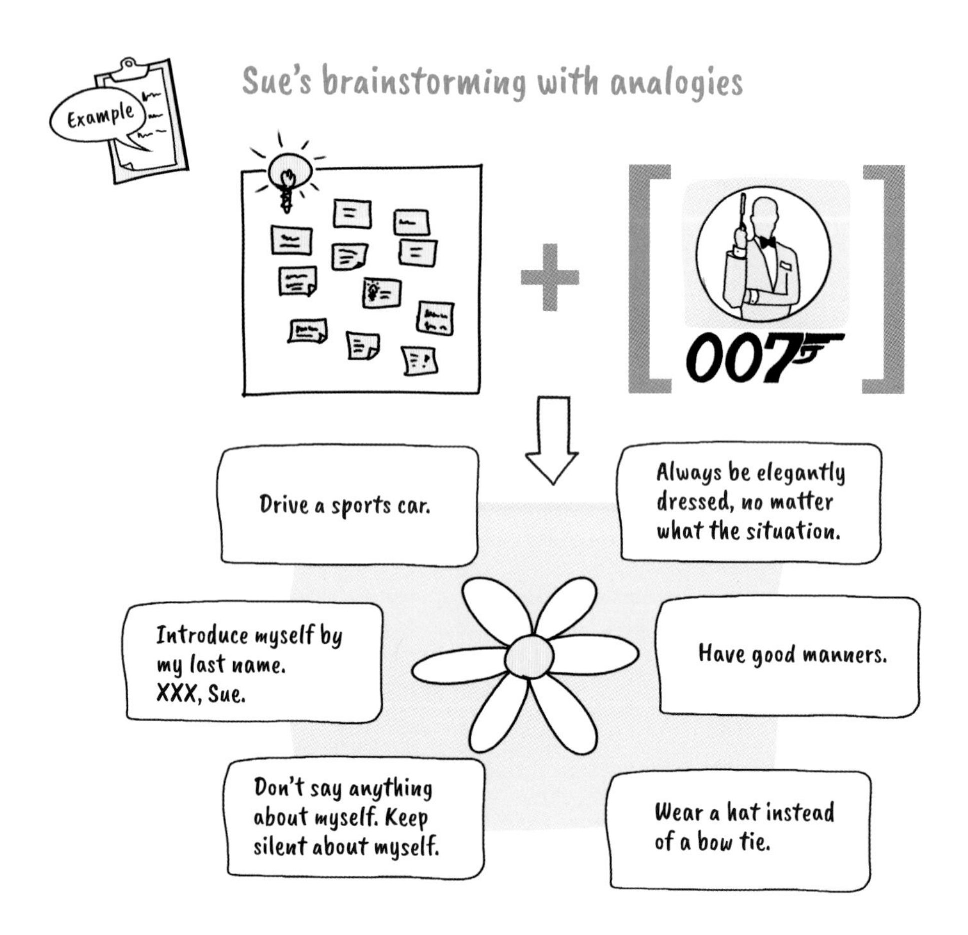

Brainwriting triggered by analogy

On the following pages, you have space for a brainwriting session on one or more analogies that you have defined on page 131. After brainstorming, select ideas that suit you and enter them into a daisy.

Brainwriting triggered by analogy

Brainwriting triggered by analogy

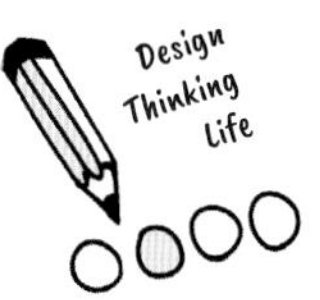

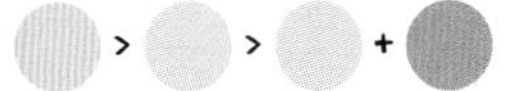

Selection of solutions

On the following pages, you have space to create more daisies and to sort your thoughts. It is important to choose the ideas from the analogy exercise that you find useful for reaching the desired stage.

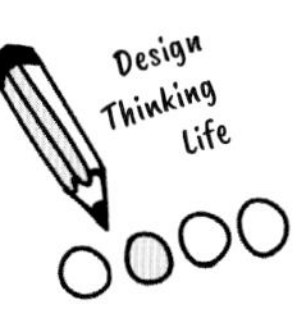
Design
Thinking
Life

Do we need more creativity?

By now, we have collected a lot of ideas, and it makes sense to conclude this phase and move on to convergence.

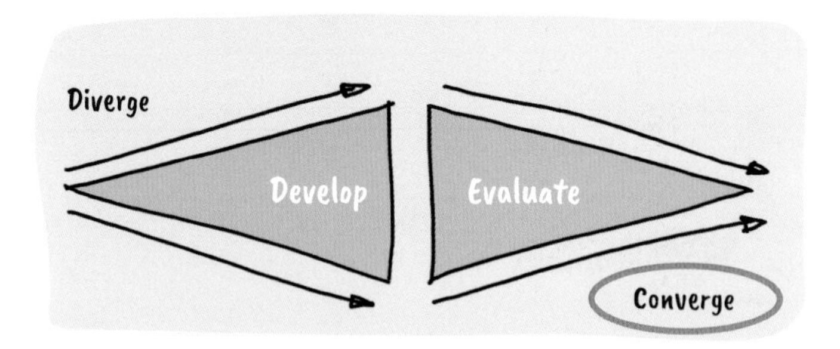

Of course, we can also use other brainstorming and brainwriting methods (e.g.,"dark horse," a technique in which we omit elementary things, as in John's case, for example: "What would life be like without my wife?"). Most of the time, however, we have already collected a lot of potential solutions for a better life at this point, so we can continue with the DTL process. That is, we focus on the solutions that are feasible and can help us feel happier. We enter the "groan zone."

Why "groan zone"?

The transition from a diverging phase to a converging phase is also known as the *groan zone*. This is because it's a challenge to select realistic possibilities for change from the multitude of ideas and put the stages in meaningful order.

In the end, we should limit our possible selection to a few approaches to a solution. The more options we have to choose from, the more difficult the selection becomes. We don't need to worry at this point about leaving out a brilliant idea. In the end, we usually decide intuitively what the most suitable selection is at the moment.

For the selection, we can also add the following four questions, which we have used previously:

- How does the solution feel?
- Which solution feels better? Where and what do you feel in your body when you imagine the solution?
- How would others know that you are trying out something new?
- What would you do more of? What are the factors interfering with a positive feeling?

In the following pages, we will perform four steps.

1. Summarize and select ideas.
2. Visualize ideas and target the state.
3. Test life concepts and experiment with them.
4. Improve the stages and the suitable life concepts iteratively.

How can we summarize and select ideas?

To select and compile ideas, we can put ideas on a so-called decision matrix. For example, we can label the axes with "cool" and "feasible." Our ideas from the different daisies can be sorted very quickly in this way.

Sue's selection of ideas

Since Sue has already developed some cool ideas, we stick to her decision matrix. She sorts the individual ideas into the grid according to her feelings. Sue wants to follow up on everything that is cool and feasible.

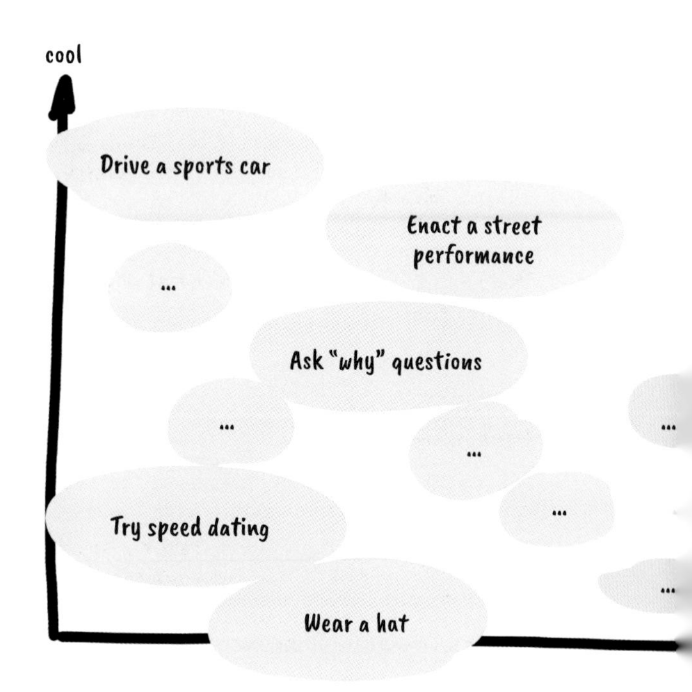

On the following pages, you can also convert your ideas into a such a grid. It usually works for desired changes in terms of career, family, relationships, and health.

Of course, you can also define other axes or further narrow them down – e.g., "affordable" instead of "feasible" – so that the matrix comes closest to meeting your personal needs.

Which ideas are cool and feasible?

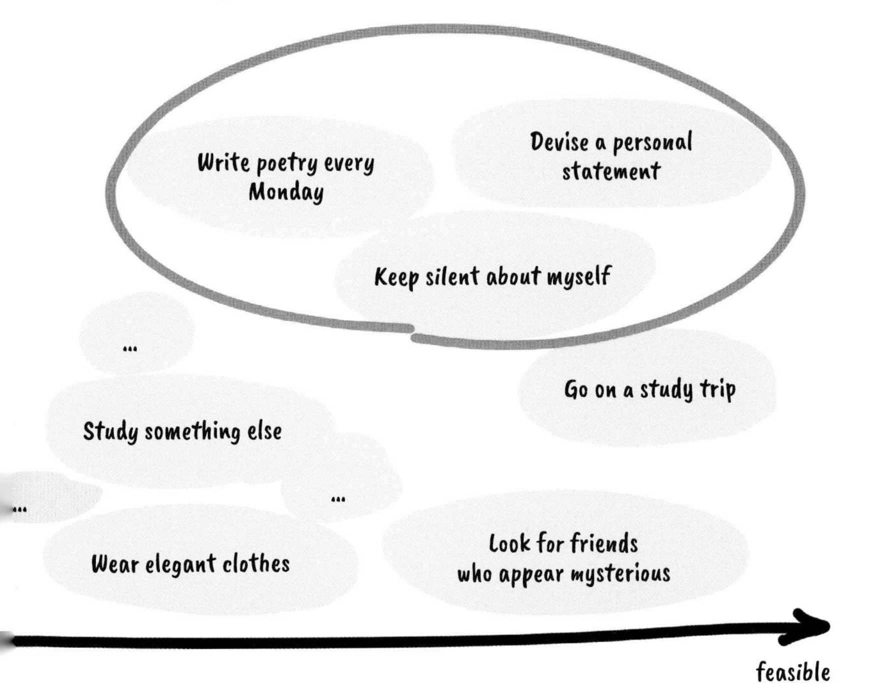

Summarize and select ideas

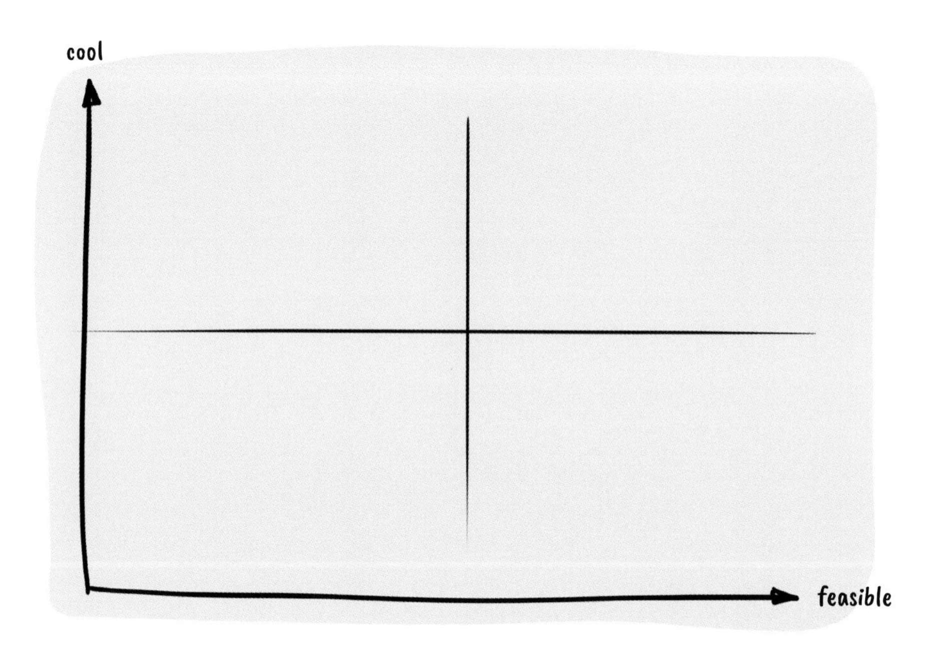

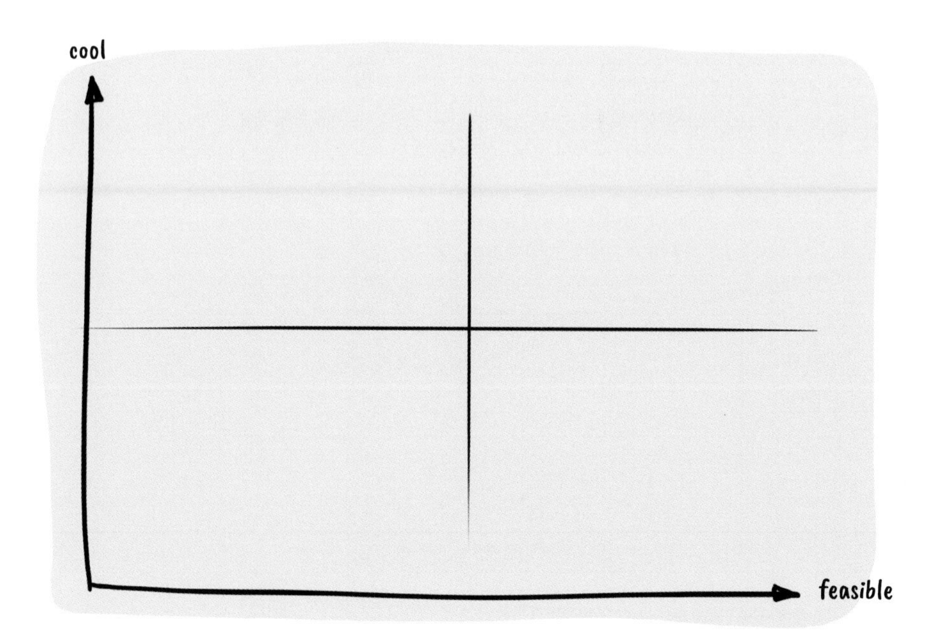

Design
Thinking
Life

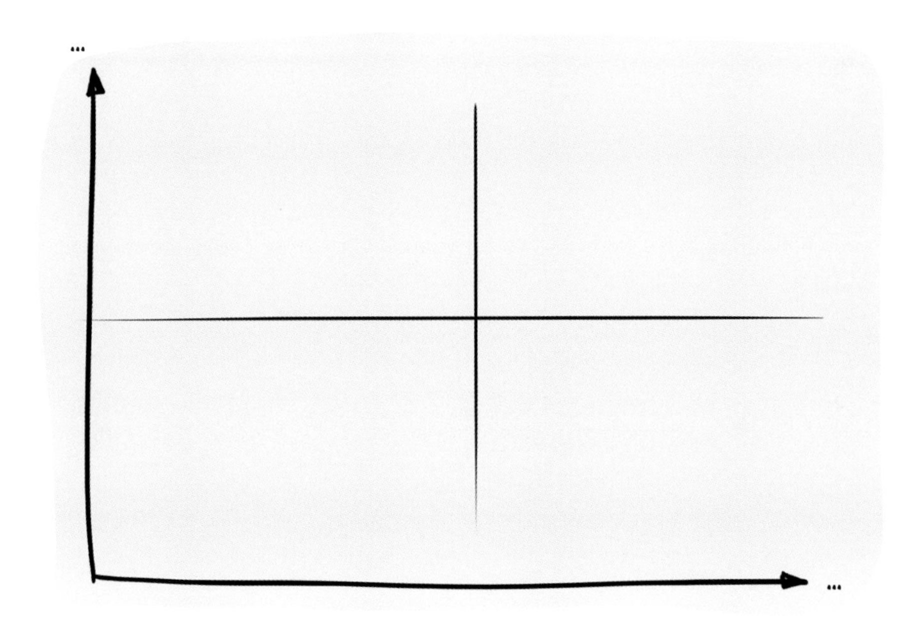
...
...

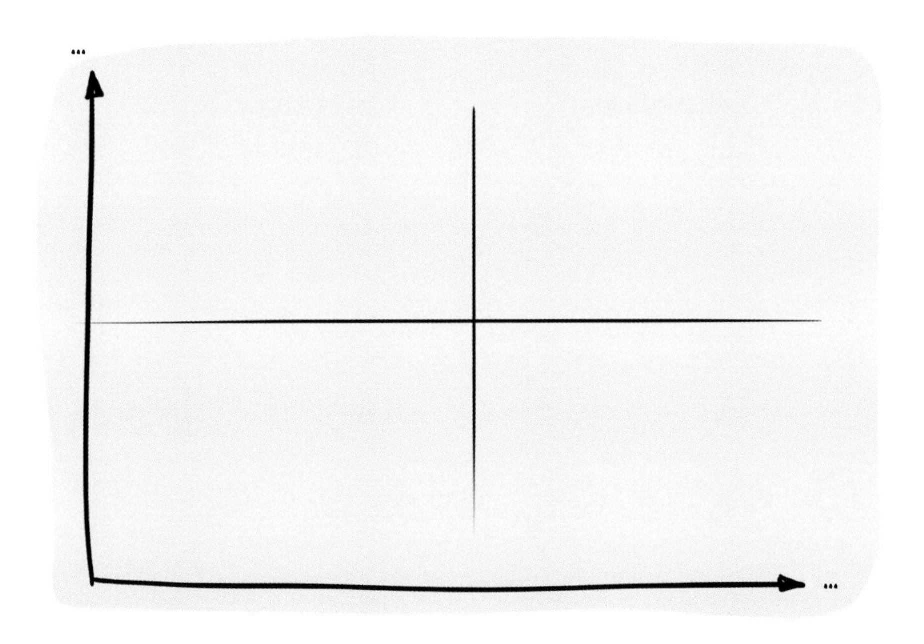
...
...

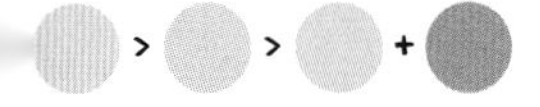

How can we make further selections?

By compiling solutions after our initial selection, it is usually possible to decide on three to six specific solutions, which we can then test.

If we are not yet sure which of the selected ideas we should experiment with, we can do a short cross-check again. It has always worked well for us to use a four-star ranking, where one star indicates that the idea is good but is not really a match for us and four stars indicate a bull's eye. We apply this ranking to three questions to obtain an overall star rating.

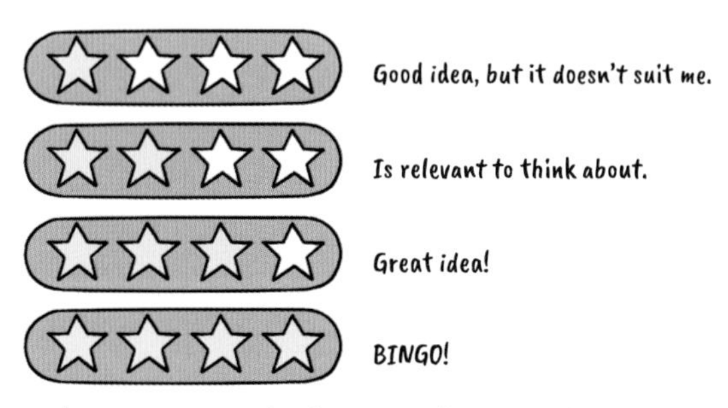

Ideas that receive more than two stars in the overall star rating are very good candidates for an experiment. This quickly gives us an idea of how we feel about the potential solutions. The following example shows how it works.

Sue's rating of "Keep silent about myself"

Sue likes all three ideas and feels they are cool and doable. However, "Keep silent about myself" is not consistent with how she imagines things to be.

I have enough self-confidence for the implementation

The idea is consistent with my values

I'm comfortable with imagining it

Therefore, she decides to experiment with something that is proactive and more in keeping with her nature. She wants to write poetry and share her poems on social media.

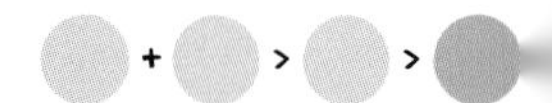

Rating of favorites

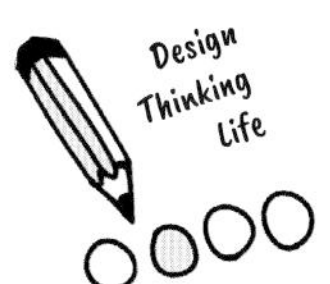

Reflect on your favorites with the four perspectives.

Rate favorite 1: . . . .

I have enough self-confidence for the implementation ☆☆☆☆

The idea is consistent with my values ☆☆☆☆

I'm comfortable with imagining it ☆☆☆☆

Overall Design Thinking Life star rating of idea ☆☆☆☆

Rate favorite 2: . . . .

I have enough self-confidence for the implementation ☆☆☆☆

The idea is consistent with my values ☆☆☆☆

I'm comfortable with imagining it ☆☆☆☆

Overall Design Thinking Life star rating of idea ☆☆☆☆

Rate favorite 3: . . . .

I have enough self-confidence for the implementation ☆☆☆☆

The idea is consistent with my values ☆☆☆☆

I'm comfortable with imagining it ☆☆☆☆

Overall Design Thinking Life star rating of the idea ☆☆☆☆

Design, test, and implement life plans

The visualizations of our potential life plans are our prototypes, which we design into a solution on the basis of the selected ideas. However, these ideas are nothing more than hypotheses that hang in the air until we have tested them in real life. Practical experience and interactions with other people in the context of our prototypes are crucial for getting feedback and ultimately designing a life iteratively that makes us happier. Whether the respective hypotheses prove to be useful, however, only becomes apparent during the implementation of our life plan.

Prototypes take the form of visualizing the stages and life concepts

For a visualization of the selected ideas and life concepts in the prototype phase, it has always worked well for us to use a grid, which begins with a starting phase, portrays three stages, and sketches an ideal outcome at the end (see Sue's life plan on page 149). This is also our first draft of a life plan, i.e., for changes we would like to test.

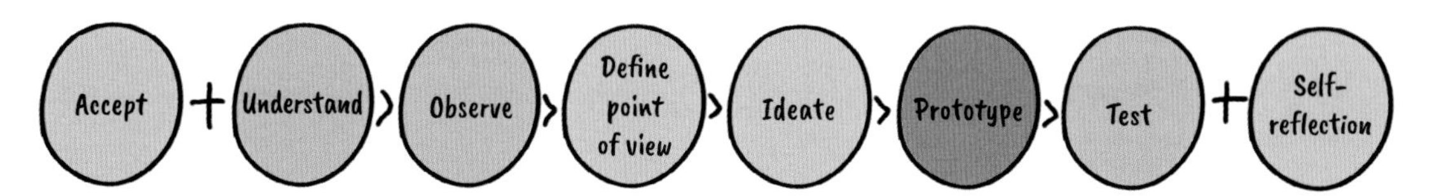

We use the grid for both large and small changes. However, in the case of major changes, the time span in which the individual stages can be tackled is usually somewhat longer.

We recommend describing several life concepts and their stages. Each plan should have a name, just as we in design thinking give each of our prototypes a name. In addition to self-reflection about the ideas already discussed, we should formulate the main questions to be answered by the life plan.

On the opposite side, Sue has already begun to visualize her life plan, while at the same time working on questions that this life plan addresses. She then reflects on how she feels about how she imagines things; whether she has enough self-confidence for implementation; whether this plan is consistent with how she imagines things; and whether she has the time, energy, and strength to tackle this change. Sue also briefly summarizes her findings again at the end.

Design thinking is about designing, experimenting with, and implementing ideas. If we separate this trinity, the end result will suffer.

Sue's draft of a life plan "More poetry in everyday life"

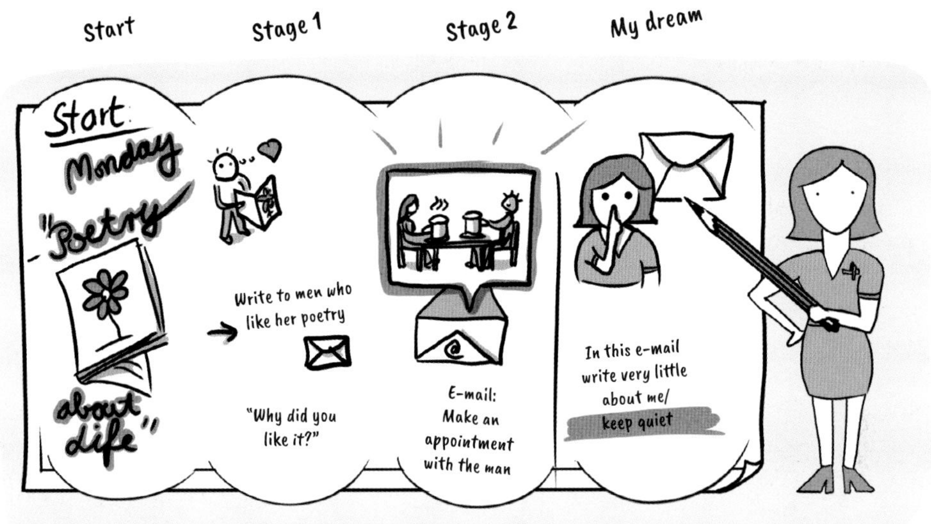

What needs does the life plan address?

- Getting more attention
- Learning a new questioning technique
- I can get to know men and appear mysterious at the same time

What questions does the life plan leave open?

- How long will it take for me to be successful?
- How can I make sure I meet someone in Hong Kong?

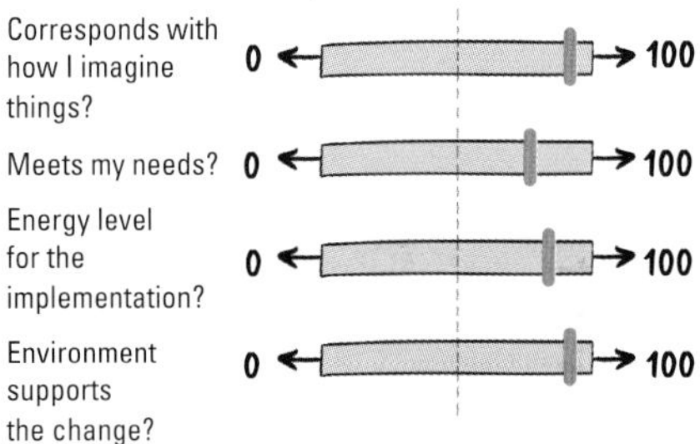

Perform a holistic view of the life plan by reflecting on whether the life plan corresponds with how you imagine things, meets your needs, and matches your current energy level for implementation. In addition, check whether your environment would support you in the change.

On the following pages, you will have space to put your plans and stages in tangible terms. Based on the new ideas, the path to the ideal outcome might change because you have learned more about your needs during the last few days and months.

Sketch plans and stages

Life plan 1: . . . .

Start:

Stage 1:

Stage 2:

What needs does the life plan address?

. . . .

. . . .

. . . .

What questions does the life plan leave open?

. . . .

. . . .

. . . .

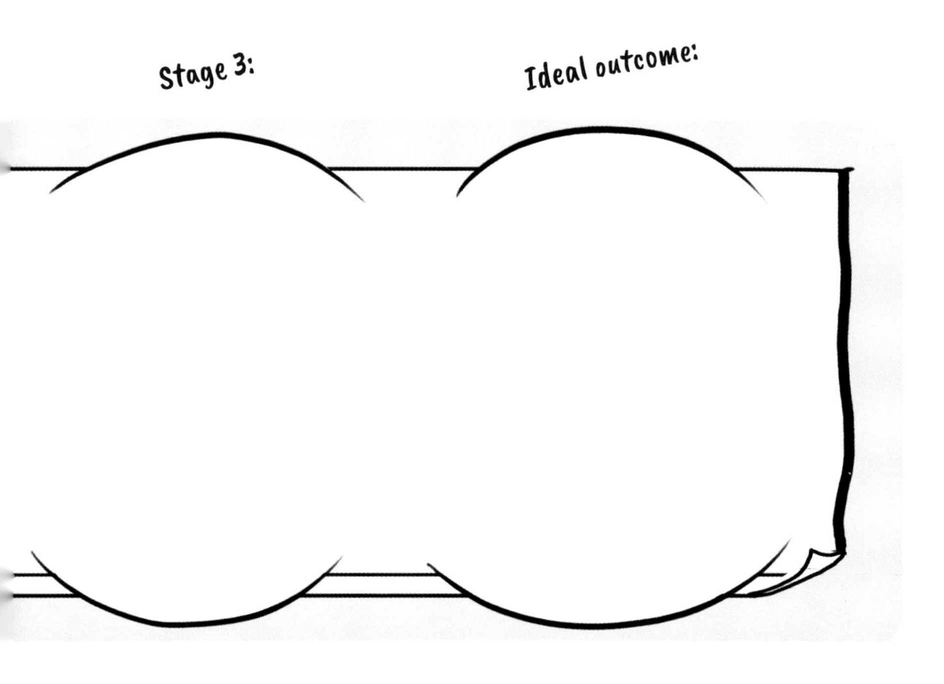

Holistic view of the life plan

Corresponds to how you imagine things?	0		100
Meets your needs?	0		100
Energy level for the implementation?	0		100
Environment supports the change?	0		100

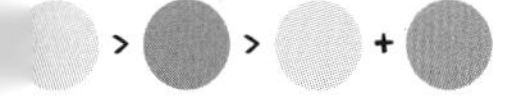

Sketch plans and stages

Life plan 2: . . . .

Start: **Stage 1:** **Stage 2:**

What needs does the life plan address?

. . . .

. . . .

. . . .

What questions does the life plan leave open?

. . . .

. . . .

. . . .

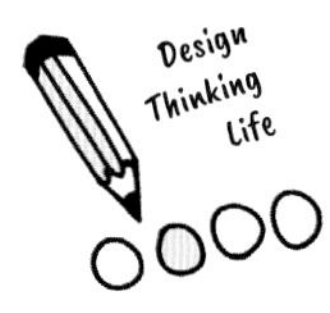

Stage 3:

Ideal outcome:

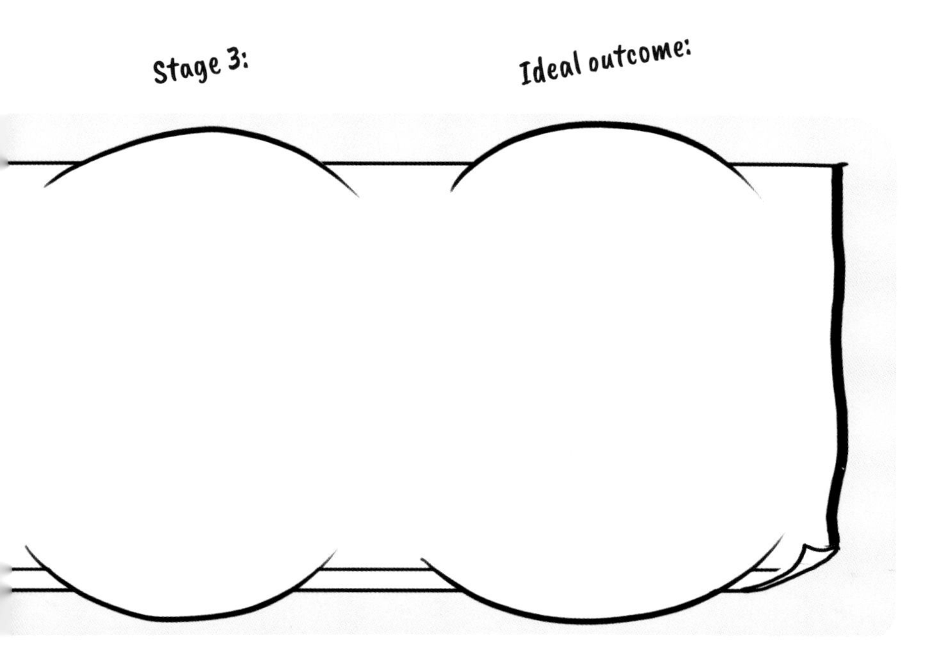

Holistic view of the life plan

Corresponds to how you imagine things?	0 ← → 100
Meets your needs?	0 ← → 100
Energy level for the implementation?	0 ← → 100
Environment supports the change?	0 ← → 100

Sketch plans and stages

Life plan 3: . . . .

Start: **Stage 1:** **Stage 2:**

What needs does the life plan address?

. . . .

. . . .

. . . .

What questions does the life plan leave open?

. . . .

. . . .

. . . .

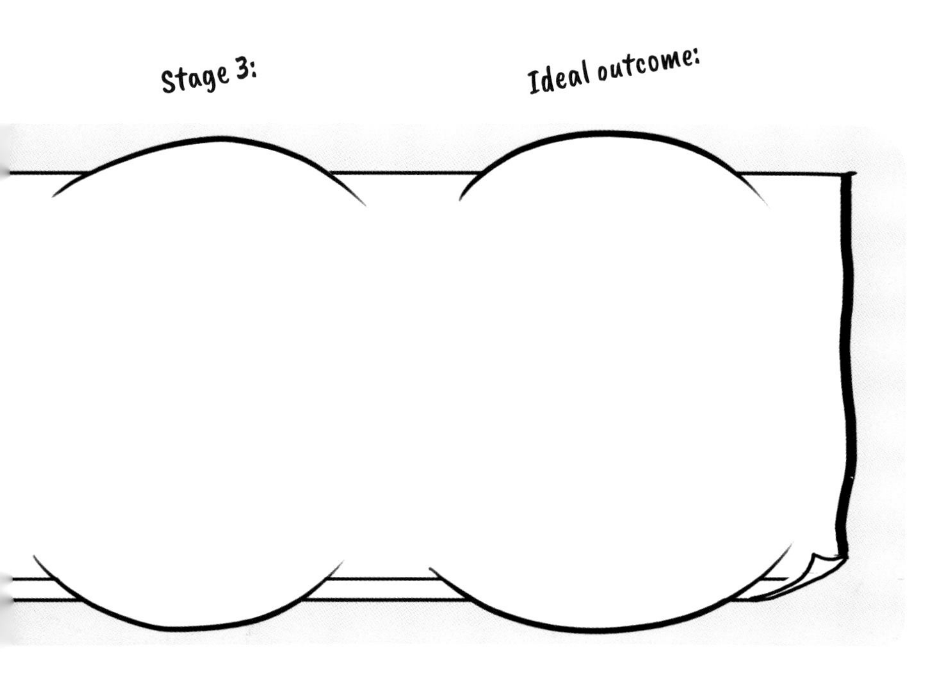

Holistic view of the life plan

Corresponds to how you imagine things?	0	←——→	100
Meets your needs?	0	←——→	100
Energy level for the implementation?	0	←——→	100
Environment supports the change?	0	←——→	100

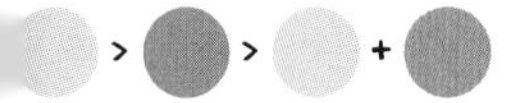

Space for notes

The "DESIGN THINKING LIFE" lives by the principle "Love it, accept it, reframe it, or trigger change." And so it is with our life concepts.

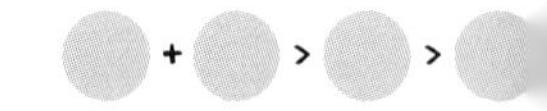

Test and experiment with the life concept

An important part of the design process is testing the idea or the plan to see how it feels and how we can improve it. If we don't like it, we can discard the underlying ideas and begin with a new idea.

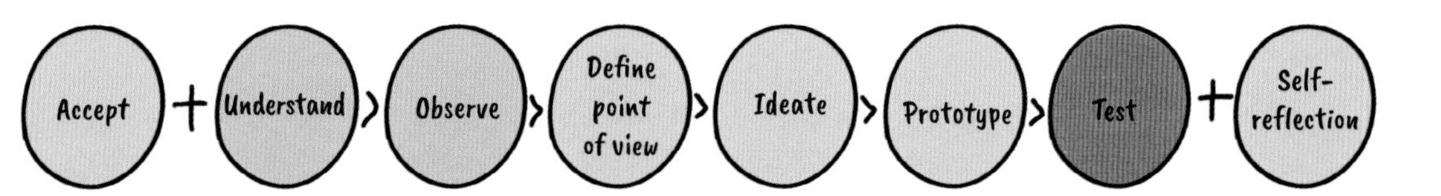

Testing a life concept is easier said than done, but there are several ways to gain insight. Testing is one of the most important steps in the design thinking cycle because it is not a matter of thought experiments but of real experiences.

Testing requires, above all, curiosity and the willingness to try something new.

It is also important to emphasize that testing is about finding out, with as little effort as possible, whether a potential life plan gives us pleasure and whether we should consider realizing it.

What possibilities can we use to test life concepts?

If we want to change professionally, we can try out a certain area for one day. If we want to live in another country, a vacation in the country where we want to emigrate is a good experiment for getting to know this country. The same is true for making a desired change in a relationship or for lifestyle changes.

What's important is that we experience individual stages or the ideal outcome and in this way gain new insights. Often, we have assumptions about what constitutes a good life or how we can be happy in a certain profession. But we will find out what it really feels like only when we experience the situation in a real environment. So we act by implementing small prototypes of stages and ideal outcomes.

None of us likes to fail, but an early failure – in terms of trying out approaches to a solution – provides valuable stimuli for change.

Our prototypes in the form of life plans and stages can be tested and validated in very different ways. The goal is to find out what makes you happy.

Decreasing time spent at your current job and testing another life concept for a while is costly and involved, but is the best way to gain real experience. “Getting a taste” in the context of a one-week internship also yields important insights; so do forms of participatory observation (so-called shadowing) of life concepts, professions, and activities.

A simple form of shadowing is to talk about your ideas and share them with others. This includes interviewing experts, as well as people who do what you want to do, or talking with people you are close to, such as friends and family members. You have countless people in your network with whom you can share and discuss your ideas.

How Sue tests her prototypes

Sue has discussed her idea of a recurring social media post on the subject of poetry with her friends. Most responded positively and encouraged her to create a post. Many also offered to "like" the post, in order to spread it virally and attract attention.

As long as we can learn from our prototypes of life concepts, it makes sense to experiment with them.

In addition, there are various other possibilities – such as lectures, field reports, or blogs – to learn more about specific topics you would like to change about yourself. Beyond that, courses are offered today on the "DESIGN THINKING LIFE" or under the slogan "DESIGN YOUR FUTURE" in nearly all cities and online, which provide an opportunity to enter into an extensive discussion with other participants, receive feedback, or simply let yourself be inspired.

Feedback capture grid

Either way, we get valuable feedback when testing our ideas, stages, and aspirations in life. A simple matrix with four quadrants in which we write down the feedback is suitable for capturing feedback. This gives us a good overview of what we thought was great, what we will discard, what has no relevance to our aspirations in life, and, of course, new things we have experienced and want to pursue.

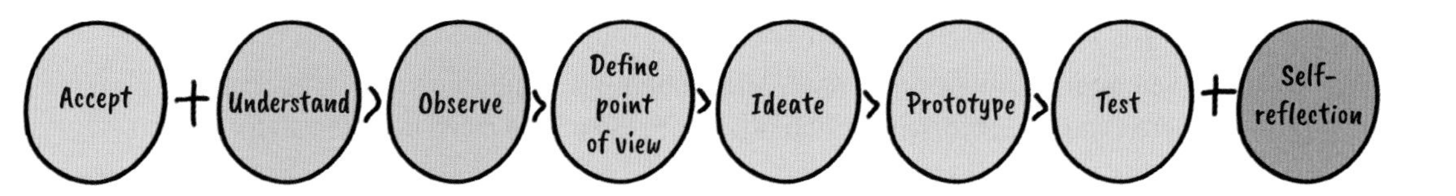

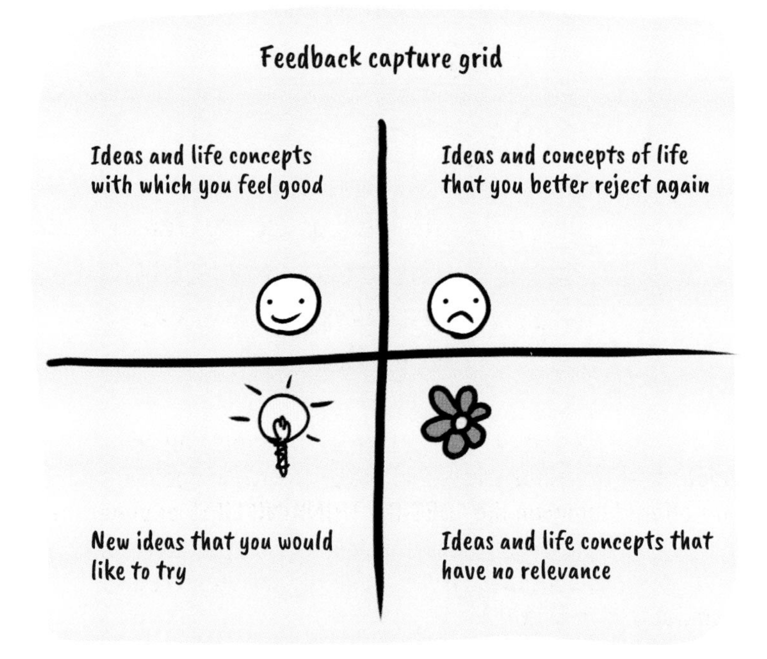

Because design thinking is an iterative process, it is advisable to repeat the previous phases a few times and reflect on them accordingly to improve, adapt, or discard our ideas, stages, and life concepts.

When we have finally drawn up a realistic life plan, it is valuable to describe or visualize it in a comprehensible story. A sketched story has the advantage that each stage can be represented as a picture, expressing the individual changes we imagine.

John's life plan: Sketch of his self-efficacious action

John's expectation of self-efficacy, for example, now stands in a new relationship to his own desires, his skills, and the actions he himself can carry out successfully. However, we will also find that this story will change over time because many changes also depend on the social system. Here there are interrelations and occasions for feedback that make our life a complex undertaking.

Capture feedback

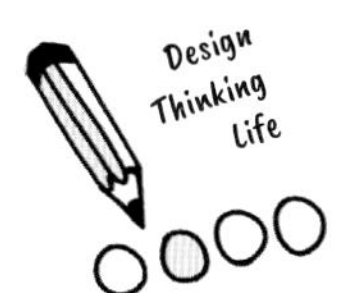

On the following pages, you have five empty feedback capture grids at your disposal. These grids give you enough space to document the important findings of individual experiences.

Feedback capture grid

Design
Thinking
Life

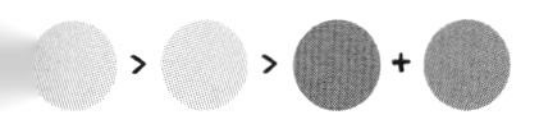

The feedback from the experiments with stages and life plans helps you fine-tune the existing plan, to include new elements or to formulate and visualize a new life plan or individual stages. You can do this iteration (test – obtain feedback – improve) as often as you want until you have found a concept that fits your individual needs.

Draw up your life plan and your self-efficacious way of acting:

Design
Thinking
Life

Why DTL is a team sport

Throughout the DTL cycle, it is important to involve our network, our friends, and our family. We can use them in different ways in our DTL efforts. You may have already involved these people at different points. For example, you may have interviewed them during the course of exploring your self-image and others' perceptions of you, or you may have discussed stages and life plans with them in order to obtain valuable feedback.

These people are particularly important when you're making a change. They can help you make contacts when you think about switching jobs. But they also listen when you have had an experience with an experiment and can motivate you to continue if you have failed or the experience was negative.

So you need these supporters, allies, and close family members to initiate change. Often, your own change has a positive impact on others, who feel inspired by it and find the courage to change something in their lives, too.

DTL is a team sport

Which people can give you help in the process of change and with practical implementation?

DTL is not just about changing the individual – each stage and change has an impact on many people in our environment.

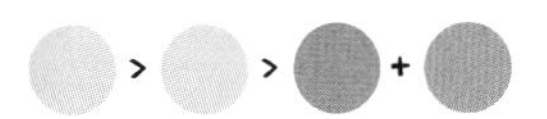

How do we implement the new life concept?

When implementing the stages, it makes sense to reflect in each case whether the overall situation has improved as a result of our change. If not, we have to adapt our activities again. There is no optimal final state in life. Even if we have the perfect job, the best relationship, and the healthiest life, there will be situations that need change or external factors that set our mobile in motion. So the DTL cycle is a continuous process in which we are constantly experimenting, thus adjusting our lives or redirecting the focus.

This DTL book has given us a wide range of strategies and techniques to continue the iterative process. With the right attitude, we'll be able to dig a little deeper with each iteration, learn more about ourselves and gradually make more changes if we realize we want to explore new avenues. Reframing has also taught us to draw a positive lesson from failure because it has merely been an experiment on the way to a satisfied life.

We feel happiest when we can continuously develop ourselves further and thus constantly balance our mobile of areas of life. It's best to take some time at regular intervals to reflect on where we stand and how things are going.

At the beginning of this *DTL Playbook* (pages 37–39), you rated different fields of action in order to assess your happiness with regard to these fields of action and to compare them with the initial state (see page 37). We recommend you fill in this grid again after reaching one or more stages and then decide which further changes make sense.

Self-reflection:
Where do you currently stand in life?

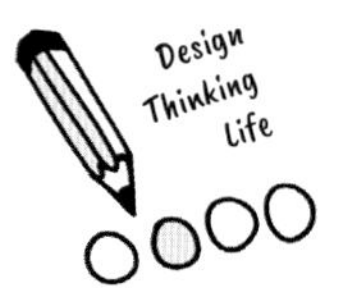

Sometimes the path reveals itself only when we start walking it.

What happened to Sue and John?

Initiate great things with small steps toward change.

Sue and John have both learned something crucial during the last few weeks and months; namely, how they can change something self-efficaciously. Both have set their systems in motion, have tried out new things, and have thus taken the change into their own hands. The most exciting thing is of course your own story because with every step you take, you write your own story. These fictional characters have also set something in motion. You want to see where they stand now.

Sue identified two big issues that she wanted to change. One was the desire to return to Europe from Asia and be closer to her family. Second, Sue felt the need to find a partner. In terms of her desire to come back to Europe, it turned out to be not Switzerland but instead Paris. And Sue associates the French "city of love" with only positive things anyway.

Out of Sue's desire to find a partner, she implemented her idea of posting short maxims with little pictures on social media. In the meantime, Sue has 4,320 followers who read, comment on, and forward her maxims on a weekly basis. She often links her quotes with questions, so that responses from readers are triggered. In this way, Sue has already made contact with several men and has had two enjoyable dates. She takes so much pleasure in writing the statements that she is considering publishing a small book titled *The Daily Poetry of Life*.

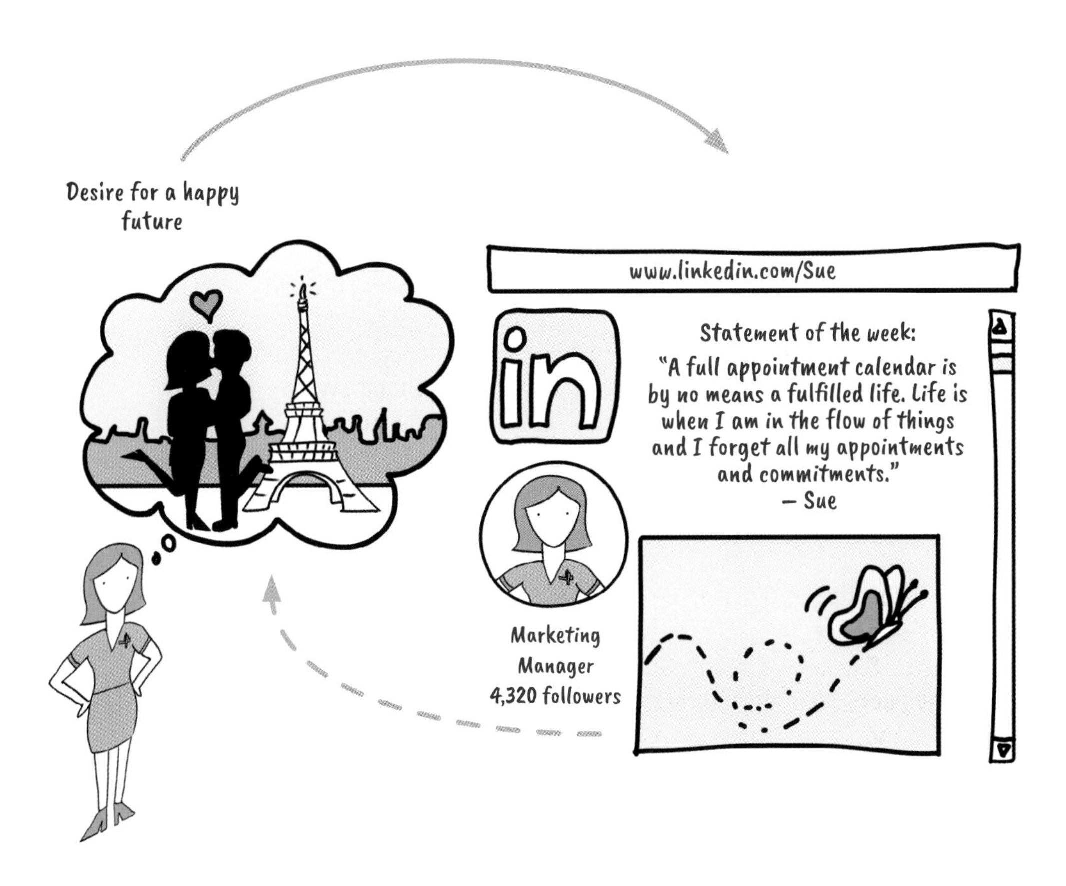

A new chapter in John's life has dawned. Being able to retire early at 60 was a huge gift of time and quality of life. However, a few issues soon caught up with him that had long been on his agenda but he never tackled due to lack of time. During his career, John could always find excuses to avoid his in-laws. Now John has to face this challenge. He, too, has gained various insights with the techniques from the *Design Thinking Life Playbook* and has become aware that he needs to start with himself in order to improve his relationship with his in-laws and Michelle.

John started with small stages and gradually tried out different things until his new everyday life felt good for the moment. Together with Michelle, he made various arrangements and came to terms with her. For example, they visit his in-laws together only every two weeks, and every second time they ride the motorcycle for the trip to New Albany.

To revive the relationship with Michelle and to spend more time together as a couple, John will join Michelle in certain sports activities in the future. Michelle has gone inline skating regularly for more than twenty years. John isn't in the best physical condition to start skating himself, so Michelle gave him a motorcycle for retirement. But not just any bike! The bike has the design of a chopper, so John almost feels like he's on his Harley. He also noticed that exercise and cycling are very good for his health. In the past, he often couldn't fall asleep, but since he has been exercising and moving a bit more, his sleep problems have vanished.

These major changes need a little more courage for their realization, and the decision-making process is often more complex.

In the next part of the *DTL Playbook*, we'll focus on the big changes. These big changes are described under the topic of professional and career planning because for many of us, it's a vital issue that we have to deal with time and again. In the case of major changes, we often question the entire system. For example, we ask ourselves whether we have chosen the right course of study or the right profession. Or, as in Steve's case, whether he should continue his course of study or work at a start-up. Such a decision has far-reaching consequences for our lives. However, it is often difficult for us to leave a certain field of activity (= frame).

Until now, we have mainly made changes *in* the system, such as using an energy journal to shift our activities around. In the case of major changes, we often work *on* the system, i.e., we make drastic decisions that are far-reaching and often completely rearrange the system.

For a better understanding, we can imagine the different types of changes in two different loops. They are sometimes referred to as "single loops" for small changes and "double loops" for big changes.

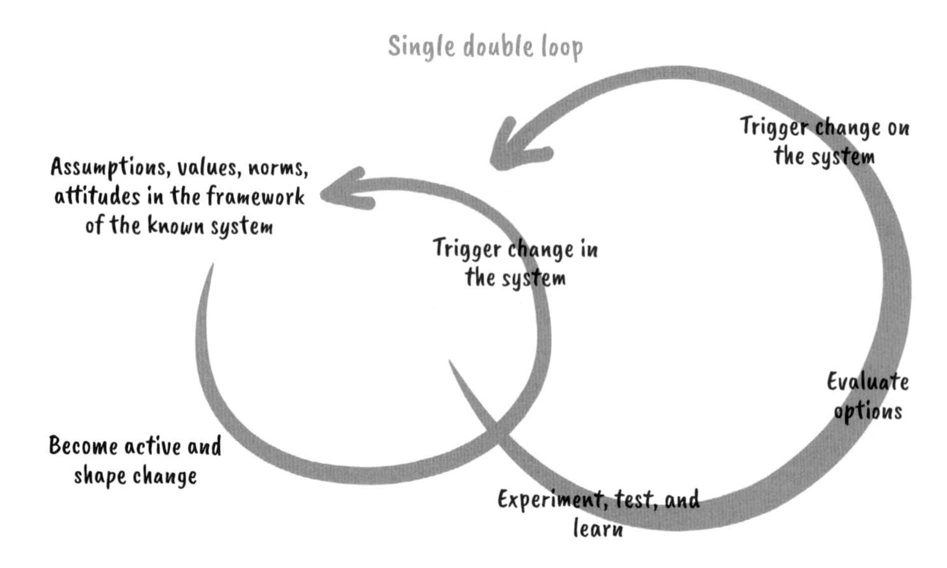

How do we know if we are dealing with big or small changes?

As we can all imagine, the question is not very easy to answer. Let's take the case of a student who is getting bad grades in his business management studies. This may be because he starts preparing too late for the exams or because his exam preparations are insufficient. In this case, it would be necessary for him to adapt his studying strategy. For example, he could study more efficiently by using appropriate time management or project management and thus achieve better results.

It might also be, however, that he's not motivated and that a closer look reveals he doesn't even like the course of study. He may have chosen business management only to avoid disappointing his parents, who expect him to take over their company later. In this case, small changes would not provide a solution. The psychological strain would remain – or even increase! – despite time management.

So we are often faced with complex changes and questions in the sense of a conflict between what we want and what would be reasonable/what is expected of us from outside.

A good way to recognize this complexity is to highlight and analyze partial aspects. At the end, we take a step back and look at the situation from a distance.

It's similar to an artist working on a painting. He concentrates on the design of individual elements, such as the drawing of the eyes in a portrait. From time to time, however, he looks at the work from a distance in order to see it in its overall context.

"In the draft, you can see the talent, in the execution the art."

– Marie von Ebner-Eschenbach

In the next section, we'll proceed in a similar way. We will find out where our thinking preferences lie, which values are important to us, which tasks lie before us, and finally what options for action we have. We work on the double loop with the aim of creating a picture from all the pieces of the jigsaw puzzle, which gives us clues for making a particular change.

Self-check

Through the strategies and techniques presented in this *DTL Playbook*, we have learned, above all, to create a new mindset that enables us to initiate change in a self-efficacious way. At the end of Part I, we would like to provide you with a little self-check that quickly shows you where you're currently standing mentally: what you have already internalized and where you can evolve in the future. If you cannot yet positively affirm many statements, we recommend that you carry out another one to three iterations of the "DESIGN THINKING LIFE" process or reflect on individual statements and figure out why you're still waiting in line here.

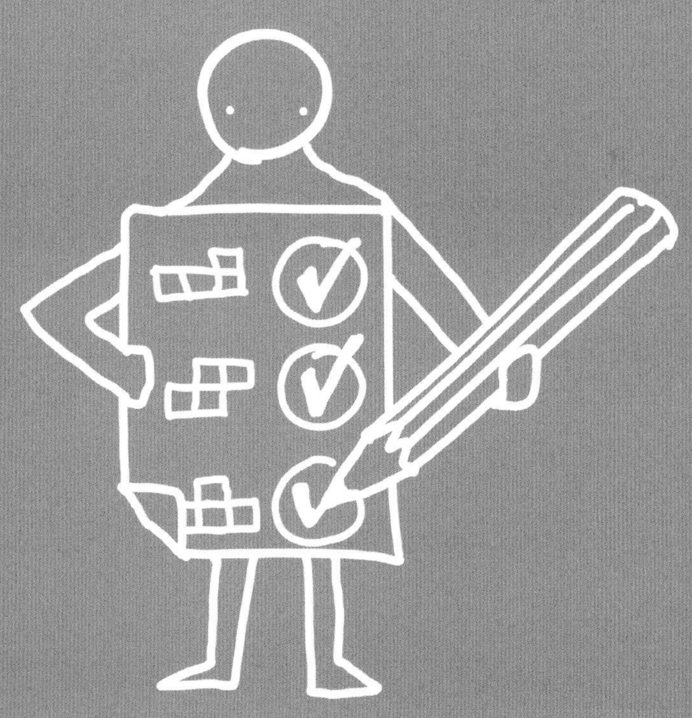

Where do you stand?

The DTL self-check helps you determine your position. It shows the depth to which you have further developed your attitude and the extent to which you are capable of acting to tackle further change.

See every single day as a possibility to grow.

Difficulties? Hard times? Problems? You don't sit around feeling sorry for yourself but actively initiate a change.

You have a clear idea of your life. You think in stages for everything that you want to change and thereby gain the ability to act.

Major events are a big challenge. You realize that you can improve your situation or look at it differently.

You don't waste energy on things you can't influence. Your focus is on doing what makes you happy.

You have your own values and conceptions that you stand for as a person. However, you are also willing to reflect on your behavior and consider the impression you make on others.

You experiment with new ideas and try to find out what the right life concept is for you. You don't throw yourself headlong into something new but, instead, iteratively explore the possibilities that should make you happier in the future.

You focus on the here and now. You use the technique of reframing to put the unchangeable past and fears about an uncertain future into another context.

You take responsibility for your own actions. If you have made a mistake, then you learn from it and do things differently next time.

You don't compare yourself with others but put your performance in the context of your own previous successes. You set yourself clear goals that you can achieve.

If you realize in an experiment that you don't like a life plan, then you change the plan or your attitude toward it. If this is not possible, you initiate a change. Act according to the motto: "Love it, accept it, reframe it, or trigger a change."

You consciously take some time for yourself, time in which you reflect on, plan, and prepare the next steps of change. You have enough self-confidence to be alone with yourself.

You know your abilities and how to use them successfully. You actively look for areas where you can contribute what you do best.

To achieve great goals and desires, you need perseverance and stamina. You muster the necessary patience until the change bears fruit and you are happy over partial successes.

You pay attention to what you spend your energy on. You don't waste time on circumstances that rob you of strength and take away the joy of life.

You stick to what's positive and get something out of every situation you encounter (good or bad).

You don't let yourself be dominated by an uncomfortable situation. You have developed strategies to recognize such situations and find a healthy way of dealing with them.

You regularly reflect on the progress you have made. You think about the good things you have achieved and correct your course where necessary to achieve your goals.

Part II

Professional and career planning

In the second part of the *DTL Playbook*, we focus predominantly on professional and career planning. Although we spend only between 50,000 and 150,000 hours of our lives working — that's just 10% to 15% of our lifetime — our job is a vital issue we have to deal with throughout our lives. Changes that lead us away from the beaten track require more courage in their implementation, and decision-making is often complex. A good way to get this complexity under control is to highlight and weigh partial aspects. So in this section, we'll explore our skills, values, and environmental factors and design and test options for our career.

Initiating big changes

At this point, you may ask why up till now we have focused only on small stages of change and haven't tackled the big issues. The answer is quite simple: Based on our experience, big changes require a lot of strength and energy. If we manage to be self-efficacious on a small scale, then we automatically have more power for big changes. In addition, it is often the case that small changes trigger great things, referring to the metaphor of the mobile again.

In addition, we have barely mentioned Steve, who – as we can guess – is faced with major decisions. He wonders, "Studying or career?" For the other personas, it is the desire to emigrate or embark on a new career that lies far outside their comfort zone. Such questions usually arise every five to ten years, and those kinds of decision often cause us sleepless nights.

Overcoming internal resistance and feeling capable of acting make up an essential key in enacting a positive change.

Overcoming resistance

For big changes, we can use the same strategies and techniques that we already made use of for our small changes. All that changes is the scale of the decision, the dimension of the stages and, of course, the planning horizon. When Sue writes a blog post, she triggers something. If she wants to stop, it's relatively easy to discontinue the blog. But if we are toying with the thought of emigrating to a distant country, it can be very difficult to revise this decision later. So let's cast a quick glance at the example of "emigrating" and look a little into the future.

Steve's brother Alex is in the final stages of earning his doctoral degree at Cornell University. Meanwhile, Alex has a three-month-old daughter with his wife Maren. At a finance and business forum, Alex meets a former fellow student who now lives in Singapore and who tells him that his bank is currently looking for analysts. Singapore has always been Alex's dream, and Maren would also like to leave the place where she was born, Ithaca (NY). Together, they imagine what life might be like in Singapore. But emigrating is a big decision for the young family. The project requires a lot of preparation and a lot of time in advance to organize things because various questions come to mind for the young family during the planning phase. Alex and Maren use the already familiar grid of life planning from page 148 et seq. to do long-term planning. Their first step is to obtain a work permit and save some money for the trip.

First, we ask ourselves what's really important to us before we build our lives on it.

The lifelong dream of Alex and Maren

Moving from upstate New York to Singapore and living a new, successful life in the big city

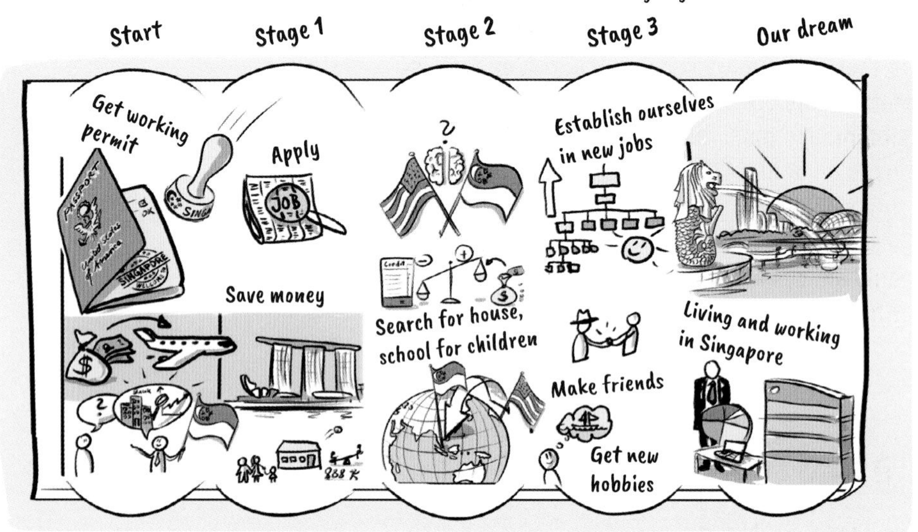

What needs does the life plan address?

- Is big-city life desirable for our little daughter?
- How easy is it for Maren to find a job in Singapore?
- Will we be able to cope with the different culture?
- Can we both have a career in Singapore?
- Can we even afford rent in Singapore?

Holistic view of the life plan

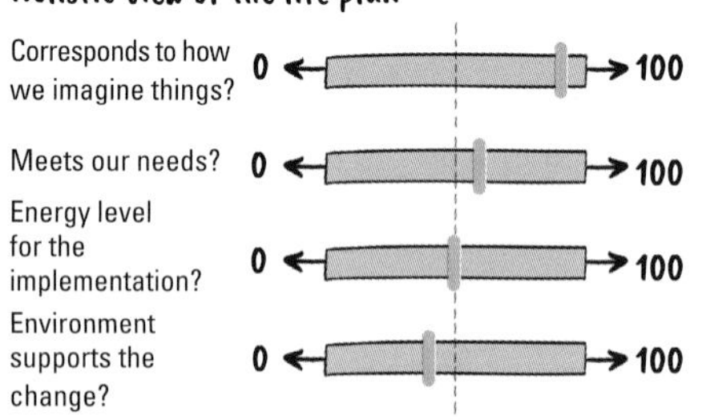

Maren and Alex get all fired up about their idea, which corresponds with what both of them imagined. Maren already sees herself shopping on Orchard Street, and Alex can totally visualize the view of East Coast Park from the apartment and the ships waiting to enter the port of Singapore. Yet they are a little afraid. Up till now, both have lived only in the United States and, apart from one semester abroad that Maren spent in Canada, have not had much experience with other cultures and ways of life. Although Alex and Steve's family originally came from China, they are the second generation living in America

To hope for change without doing something about it ourselves is like standing at the railroad station and waiting for a ship.

and are thus completely integrated into the American way of life. Asian culture is rather foreign to them. Second, they lack enough money to finance their dream. Alex has been studying for a long time, and, in recent years, he has accumulated considerable student loan debt. After evaluating the Singapore option, they realize how far-reaching this change is for them and their little daughter. Maren and Alex are also considering whether they can change their plan so that it will be less frightening to them. You can find out what Maren and Alex choose on page 242.

Life sometimes opens up unexpected possibilities. However, we should always keep the overall situation in mind to find out what our true needs are and what impact they have on our mobility.

Focus: Professional and career planning

After this excursion into the world of Alex and Maren, we will now concentrate on professional and career planning and look at a correspondingly wide horizon.

Career and professional planning is like a jigsaw puzzle. When we know all the pieces, then we only have to put them together so that a target image is created.

Other elements are also important for our professional and career planning. For example, we will find out where our thinking preferences lie, which values are important to us, which tasks lie before us, and finally what options for action we have. (For an explanation of "thinking preferences," please refer to the drawing on page 198.) All the pieces of the jigsaw puzzle create a picture that can give us clues about a specific direction to follow.

Why self-efficacy is of great importance for our career and professional life

Acting self-efficaciously plays a crucial role when it comes to tackling our goals, accomplishing tasks, and mastering challenges. People who have doubts about being able to change their own situation are more likely to be under stress and therefore more likely to feel depressed. They are less able to motivate themselves and deal less well with negative emotions.

When we come into the world, the first years of our lives are preordained. The path is shaped by our parents and their expectations, by society, and by the circumstances in which we grow up. Likewise, our school days are embedded in a rigid system in which we must function in the best possible way. Our first major decision is to choose a profession we train in or a course of study, with individual vocational schools, colleges, and universities not differing greatly in the way they teach.

However, for the first time in this phase of life, we have the opportunity to choose a path that suits our interests. It is here, though, that our first mixed feelings usually arise, and we ask ourselves whether we really want to study this or that subject. This decision will shape the rest of our lives.

The "quarter-life crisis" will then emerge, at the latest, toward the end of our studies or after our training. Suddenly, we're no longer in the protected system of an educational institution, but are called upon by the world to make something of our lives. From that moment on, we are responsible for our lives ourselves. At recurring intervals, we increasingly ask ourselves questions like:

Who am I?
What are the important things in life?
What satisfies me?

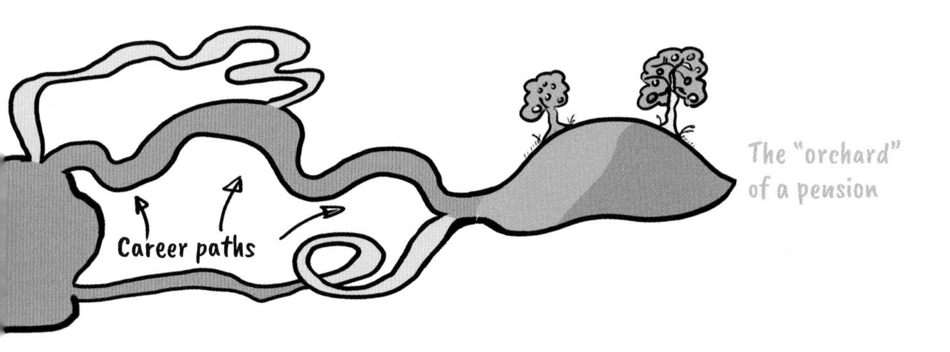

In every phase of life, we dream of a specific future, and innumerable individual longings and associated fears are added to the dream. They are often in conflict with one another, such as the desire for a stylish apartment in the city against the idea of more leisure time.

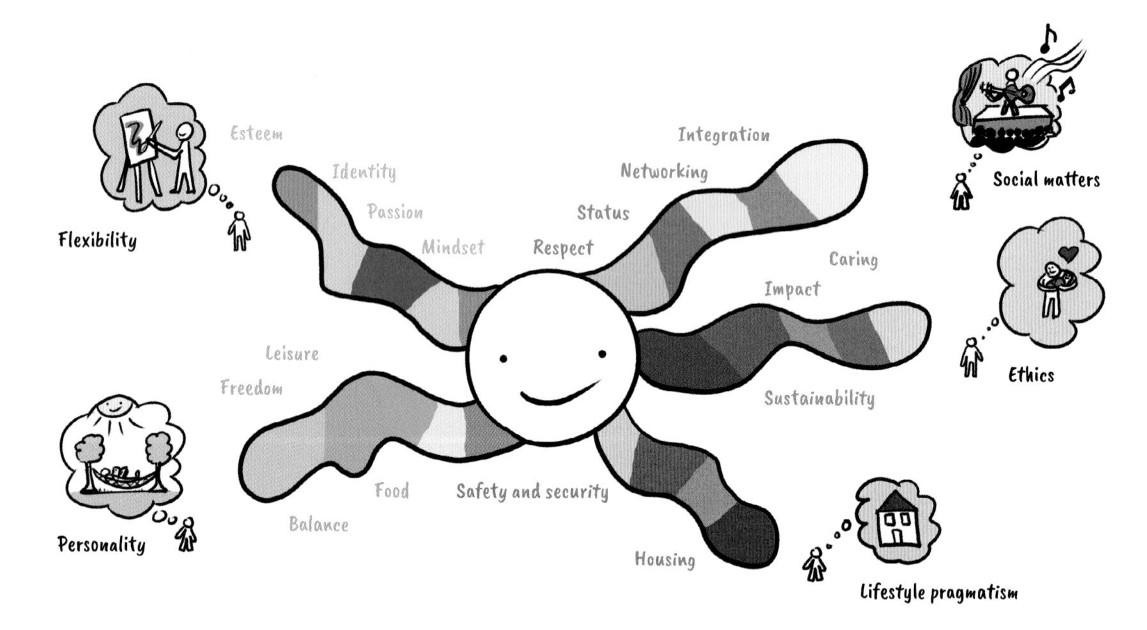

With the tools in the DTL book, we were able to heighten our awareness of which desires are strongest and have the greatest impact on our decision-making processes. To visualize the important things in life, we can once again put our desires and longings into a hierarchy.

Enter the desires that are especially important to you into the pyramid.

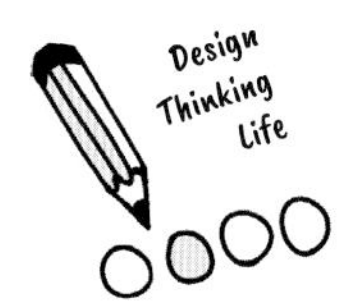

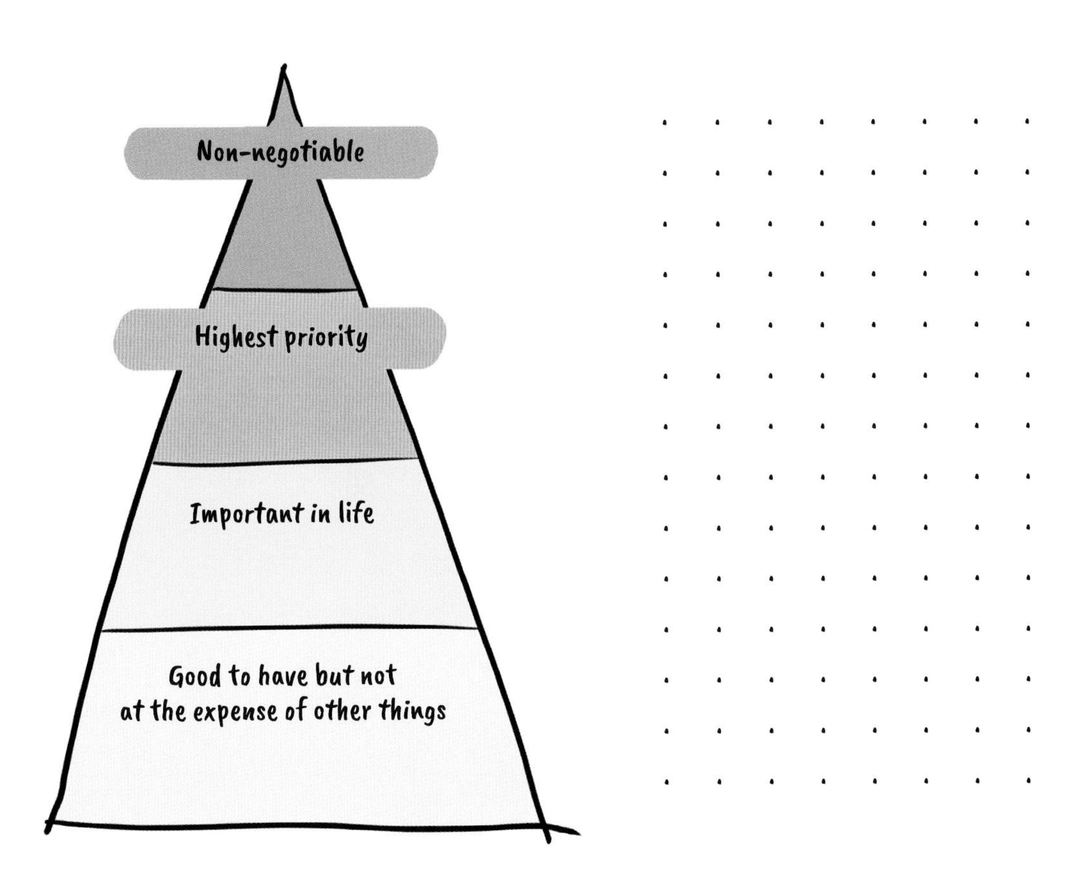

Who is Steve?

Especially when it comes to our career, there are countless possibilities; we have to make decisions, regardless of whether we are about to take the next career step, are planning to switch industries or – like Steve – are at the beginning of our career. We have to make a decision and know for sure that our next step is the right one for us. For Steve, the important thing is to decide whether he should continue his studies or enter professional life.

Steve is still at the beginning of his professional career. He has just completed his bachelor's degree in management science and engineering and is undecided about what he should do. During his studies, he was more interested in management topics than, for example, in the course on stochastics. He particularly enjoyed a one-week boot camp at the d.school, where a real problem statement could be solved using "Business Ecosystem Design." In addition to technology, the focus was on understanding complex business models and formulating a benefit that meets the needs of the customer.

In general, Steve enjoys working with other people. He is curious and likes experimenting with ideas. Most of his fellow students are rather introverted and would rather deepen their information technology (IT) skills later when finishing their master's degrees. Steve currently feels that his chosen path is more of a burden. He increasingly questions whether he should not have studied management science or intercultural communication right off, which would probably have better suited his nature.

In addition, an opportunity to join a start-up has opened up. However, the founding team is mainly looking for machine-learning experts to implement the envisaged solution. When Steve thinks about it honestly, he doesn't really want to do programming every day.

Our vocation is often at the spot where our talent is linked to the the needs of the world.

Steve has to make a decision in the next four weeks. He is still rather disoriented, and this state of suspension is reflected in his mind. More and more often he lies awake at night, brooding over his situation and thinking about his future. Moreover, the relationship with his girlfriend is undergoing a crisis. She doesn't understand his dissatisfaction. Steve's university offers "DESIGN THINKING LIFE" courses every semester. Steve would like to attend such a course and change his situation with techniques from the *DTL Playbook* to overcome his impasse and be able to act again.

Our procedural model

The journey to various career paths starts with learning more about ourselves by finding out what is important to us in life. At the beginning, we will formulate hypotheses about certain jobs, act self-efficaciously, and initiate things to find out in the end where we see our professional future.

DTL – Career Exploration Framework

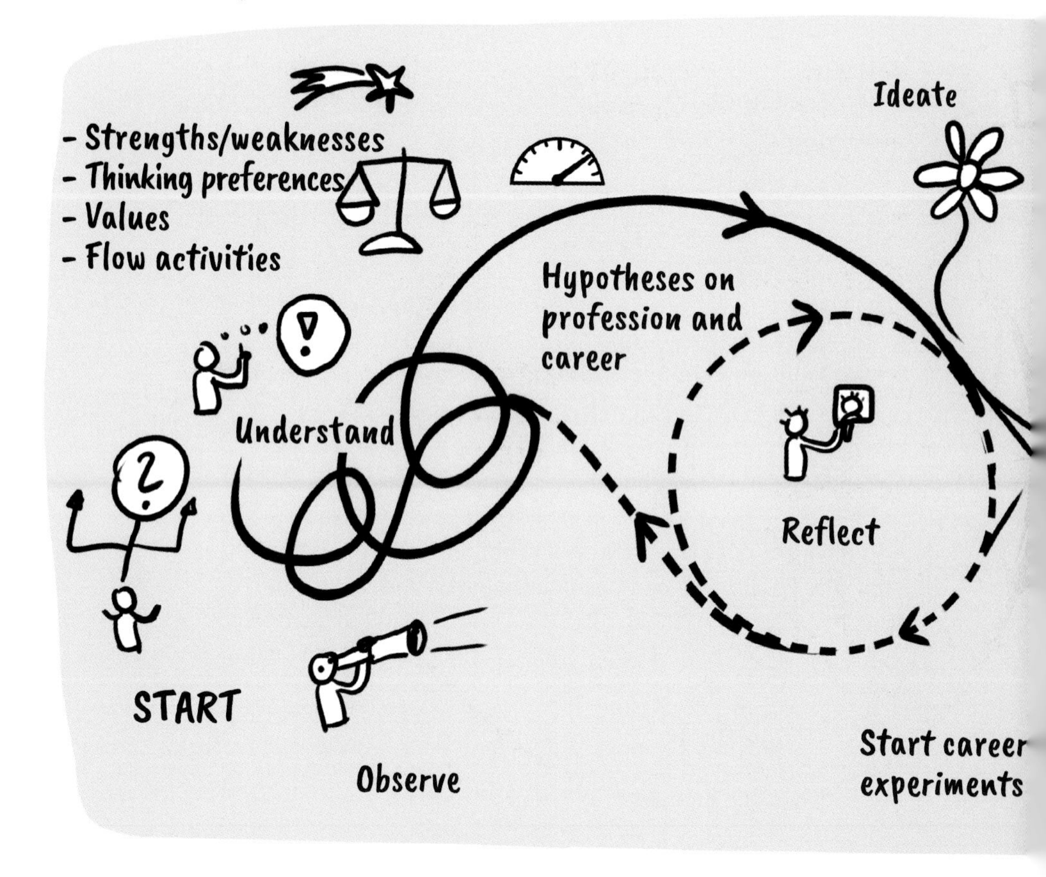

Recognize

It makes sense to test and compare different paths in tandem. We often come up with completely new ideas while testing an activity or reflecting on our situation with others. Again, it is important to proceed iteratively, to maintain the potential career paths we like, and to change or discard experiments that do not match how we imagine things.

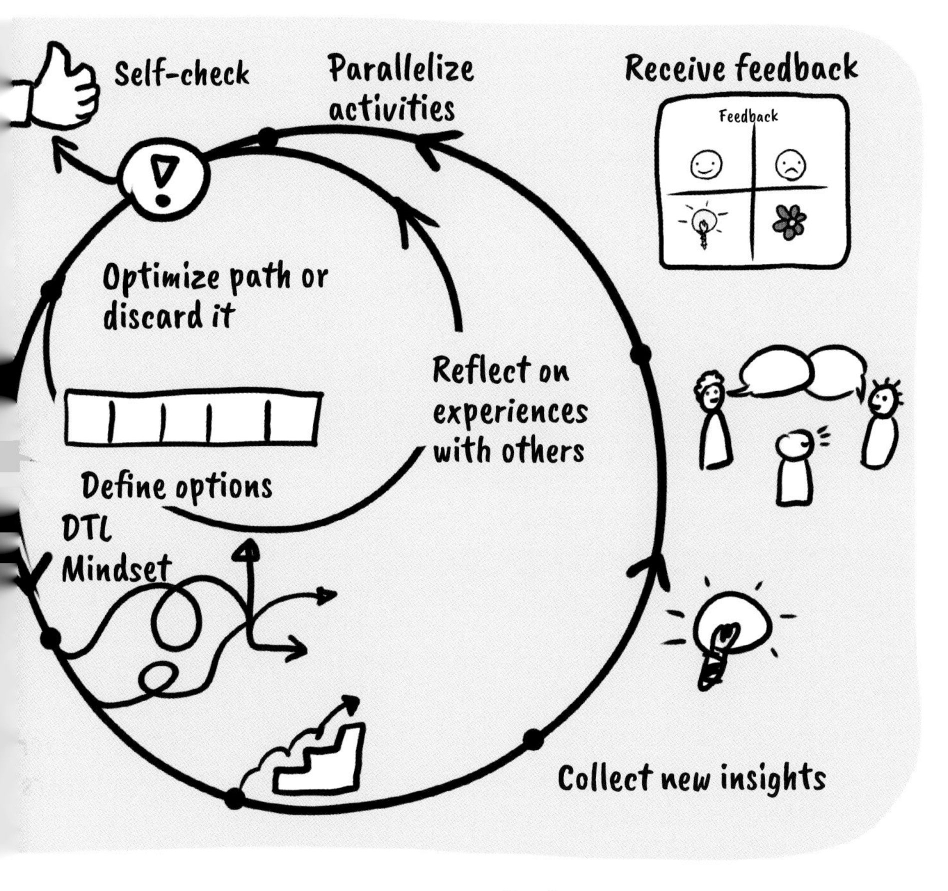

Design

Do we know our abilities, values, and motivating environmental factors?

When it comes to our career, it is important to learn more about our capabilities. In such situations, methods like the HBDI profile can help us find out where our thinking preferences lie.

The **HBDI** model (**H**errmann **B**rain **D**ominance **I**nstrument) divides our brain into four physiological sections. This model not only consists of the left and right mode but also involves the cerebral and limbic mode. The four quadrants allow us to distinguish several thinking styles. On one hand, there are the cognitive and intellectual modes, which go back to the cerebral hemisphere. On the other hand, we see the structured and emotional modes, which describe limbic preferences. In most cases, we can quickly locate ourselves on the quadrants even without the test and thus determine our thinking preference.

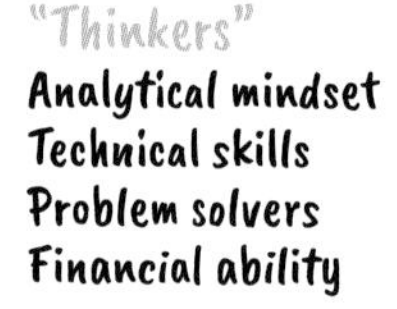

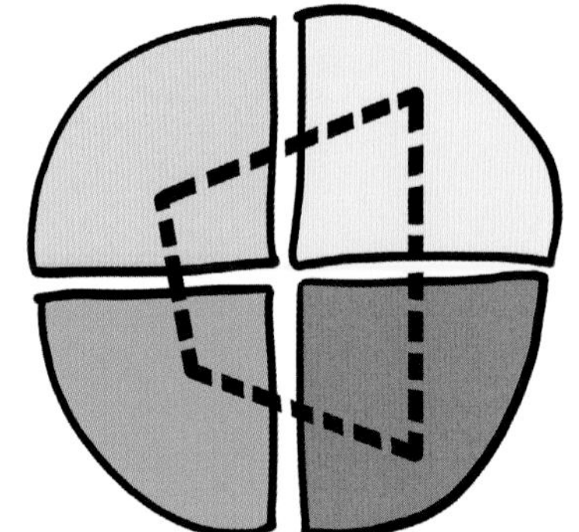

"Innovators"
Integration
Conceptualization
Creativity
Innovation

"Organizers"
Organization
Planning
Administration
Implementation

"Humanitarians"
Teaching
Writing
Communication
Interpersonal relations

A good tool for determining preferences is the following checklist, which is based on the HBDI model. The number of selected properties indicates our preferences. More information on the HBDI test can be found on the HBDI website: https://www.thinkherrmann.com.

Steve's selection of properties

[x] collect facts [ ] analyze situations rationally [ ] solve problems in a logical way [ ] argue rationally [ ] rely on numbers and values [x] understand technical connections [ ] take into account financial aspects	[x] see the big picture [ ] tolerate uncertainty and imponderabilities [x] recognize opportunities and possibilities [x] solve problems rather intuitively [x] integrate solutions and concepts [x] question known and established things [ ] use imagination and have visions
[ ] tackle problems pragmatically [ ] discover hidden problems [x] am on it, am persistent [x] manage with monitoring and in detail [ ] develop detailed plans [ ] watch out for appointments and schedules [ ] read the fine print	[x] identify interpersonal problems [x] feel reactions of other people [x] respect ethical values [ ] meet other people with warmth [ ] am enthusiastic and persuade others [x] react to body language [x] rely on my gut feeling

Steve's thinking preference is more in the yellow and red areas of the model. Had he been aware of this before beginning his studies, he might have chosen another subject that would have better suited his proclivities.

Steve's thinking preference visualized

Where does your thinking preference lie? The number of boxes checked gives you a clue.

☐ collect facts ☐ analyze situations rationally ☐ solve problems in a logical way ☐ argue rationally ☐ rely on numbers and values ☐ understand technical connections ☐ take financial aspects into account	☐ see the big picture ☐ tolerate uncertainty and imponderabilities ☐ recognize opportunities and possibilities ☐ solve problems rather intuitively ☐ integrate solutions and concepts ☐ question known and established things ☐ use imagination and have visions
☐ tackle problems pragmatically ☐ discover hidden problems ☐ am on it, am persistent ☐ manage with monitoring and in detail ☐ develop detailed plans ☐ watch out for appointments and schedules ☐ read the fine print	☐ identify interpersonal problems ☐ feel reactions of other people ☐ respect ethical values ☐ meet other people with warmth ☐ am enthusiastic and persuade others ☐ react to body language ☐ rely on my gut feeling

How does your profile look visualized? Use the number of approvals of the above assignment and draw your HBDI profile.

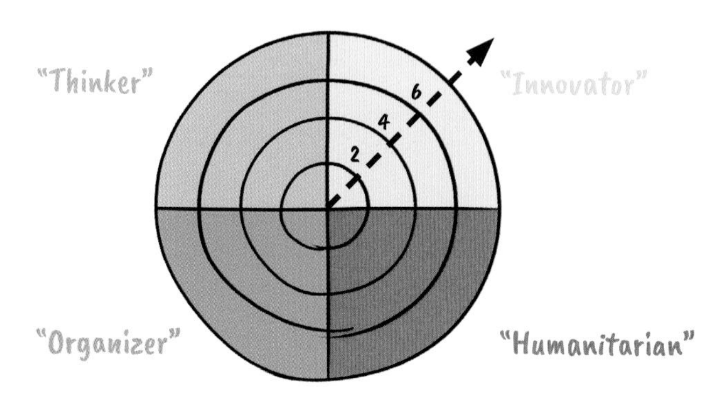

Strength = talent x (knowledge + ability)

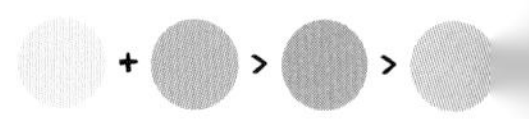

Reflect the strong manifestations, and ask yourself how you can make use of them profitably.

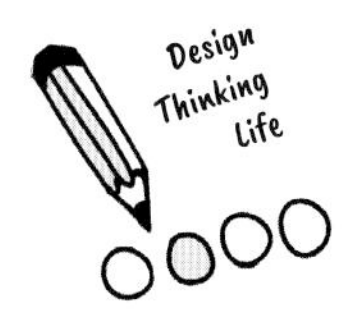

A strong manifestation in a thinking style (e.g., as an "innovator and a humanitarian" or as a "thinker")

leads to

- preferred thinking styles and talents

influences

- what interests you
- what you learn the quickest
- what inspires you
- how you communicate

leads to

- how you do what you do
- what you succeed in
- how you work with others

means to you personally?

Which values are important to us?

Other factors, such as values, can also play an important role in our choosing the right profession and industry. A simple exercise from coaching consists of reflecting on various attributes that describe values and ideas that are particularly important to us in life.

We can either create our own collection of values or make it a little easier and use the following list for a value pyramid. We select our 10 most important values (from more than 60 keywords) and prioritize them by entering them hierarchically into the pyramid. In this reflection, we consciously want to create a sequence.

Recognition	Work	Aesthetics	Appearance
Autonomy	Prominence	Profession	Be better
Movement	Relationship	Belonging	Honor
Honesty	Property	Personal responsibility	Influence
Relaxation	Development	Success	Nutrition
Freedom	Fairness	Family	Figure
Leadership	Joy	Friendship	Peace
Feelings	Caring	Holistic nature	Feeling of security
Belief	Equality	Justice	Health
Idealism	Humor	Harmony	Intuition
Creativity	Sensuality	Children	Activity
Love	Art	Standard of living	Performance
Meditation	Loyalty	Pleasure	Power
Courage	Compassion	Curiosity	Music
Order	Thinking	Politics	Openness
Travel	Partnership	Tranquility	Wealth
Self-realization	Romance	Safety and security	Self-sufficiency
Solidarity	Self-esteem	Saving	Quest for meaning
Spontaneity	Excitement	Status	Spirituality
Environment	Sports	Trust	Dreams
Knowledge	Independence	Affluence	Truth

Enter those values into the pyramid that are especially important to you. Start with the most important value.

most important

- How do you feel about your pyramid?
- For example, is wealth really your most important value, or are partnership and friendship more important to you?

In the second pyramid, there is again the option to sort your values.

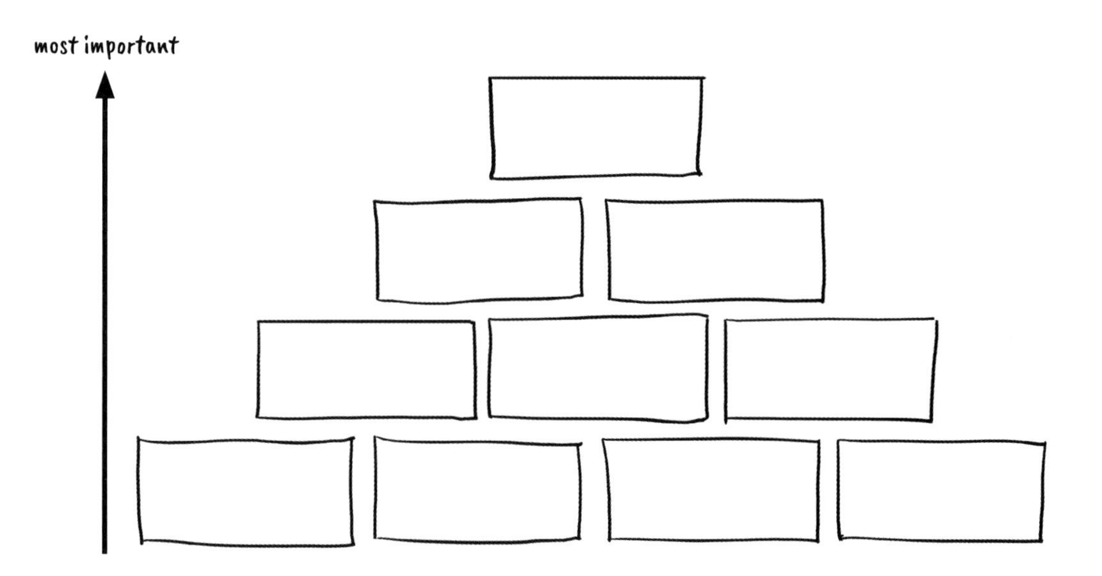

What kind of tasks do we enjoy? Which environmental factors are important to us?

The second reflection explores what kind of work and tasks we do best. We can also create our own collection of tasks and environmental factors for this self-reflection or take 10 terms from the provided list of more than 30 keywords. We can sort these factors into a pyramid to think about their importance.

- Visible results
- Freedom of choice
- Challenges
- Large projects
- Independent action
- Little control by the environment
- Little detail work
- Clear objectives
- Clear rules
- Explanation of changes
- Recognition of performance
- Clear job description
- Opportunity to ask questions
- Tasks that require precision
- Undisturbed working environment
- Variety
- Time to enjoy life
- Freedom and latitude without problems
- Flexible conditions
- Possibility to communicate
- Public recognition
- Working together with people
- Friendly, open atmosphere
- Security and stability
- Time to adjust to changes
- Work on small teams
- Recognition
- Clearly formulated expectations
- Harmonious environment
- Structured working

Prioritize environmental factors

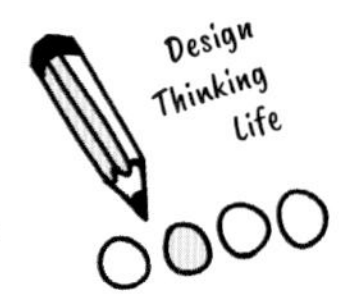

Create your pyramid with the most important motivating environmental factors.

- How do you feel about your pyramid?
- For example, are clear goals really that important to you, or is a friendly atmosphere more important?

In the second pyramid, there is again the option to sort the environmental factors.

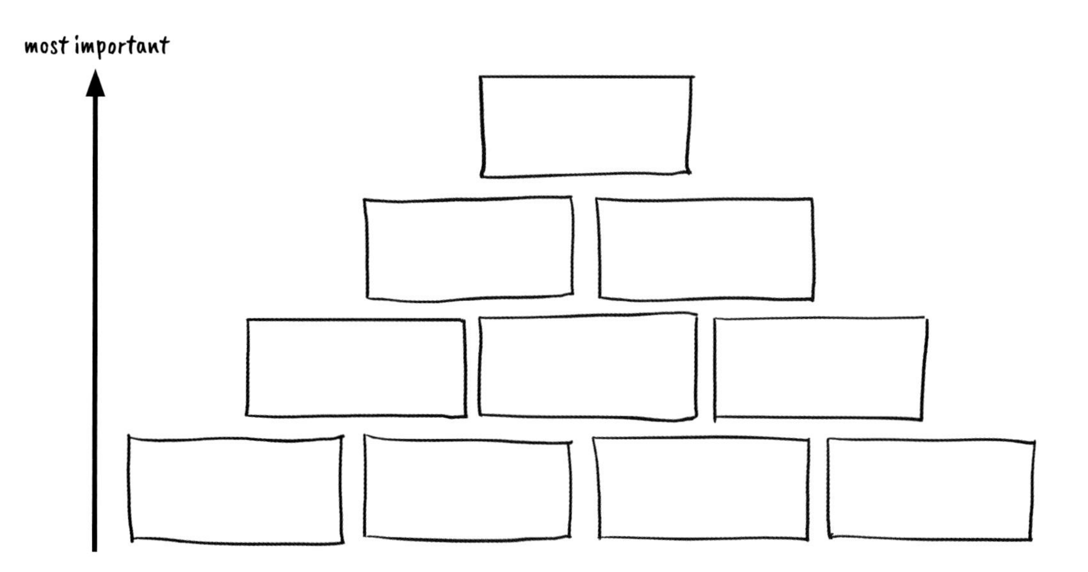

Overcome hypotheses

In addition to our values and the environmental factors, hypotheses often play a vital role in the dynamics of change. Hypotheses impact which mental model we live in, i.e., how we behave, which routines we follow, or which relationships we enter into.

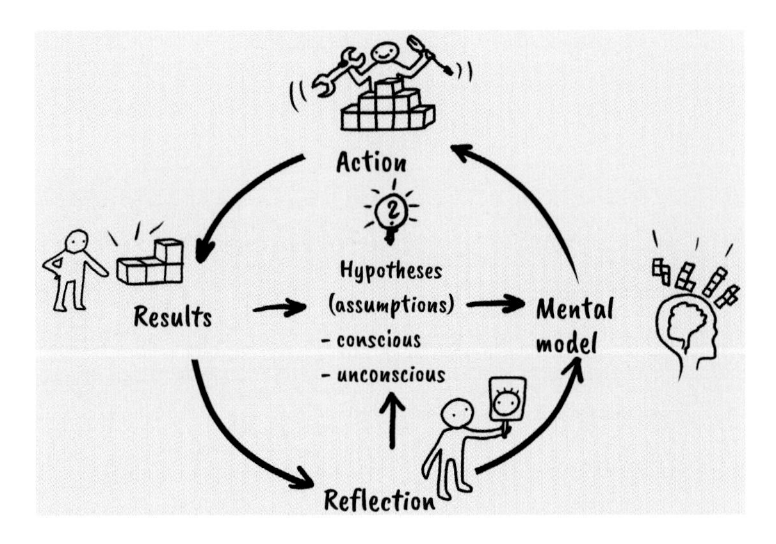

If we accept a job in a new company because we didn't like the old job, we have very specific hypotheses about what our future tasks will be, how we will work with our new colleagues, or what type of corporate culture we will encounter. And, of course, we usually assume that we will find a better situation than the one we have experienced so far. Whether these hypotheses apply or not will only become apparent in the future, however, when we have actually taken the position. The same applies to the choice of a course of study or – as in Steve's example – determining the next step after graduating with a bachelor's degree. As you can imagine, hypotheses sometimes apply, but, as often as not, hypotheses are refuted.

The challenge is that we are usually unaware of these hypotheses and therefore do not include them in our considerations. They are entrenched in us because they have worked in the past. In other words, we had a positive experience that misled us to avoid thinking about our assumptions. Often we limit ourselves to a few assumptions and disregard many others that would have been equally important and sometimes even more important. They are our so-called blind spots.

We humans usually hold on to our (conscious or unconscious) hypotheses for a very long time. It is important to make ourselves aware that they are just assumptions. From our own experience, we know that we begin to question the usefulness of these assumptions only when they prove to be unhelpful, i.e., when we find ourselves in a crisis.

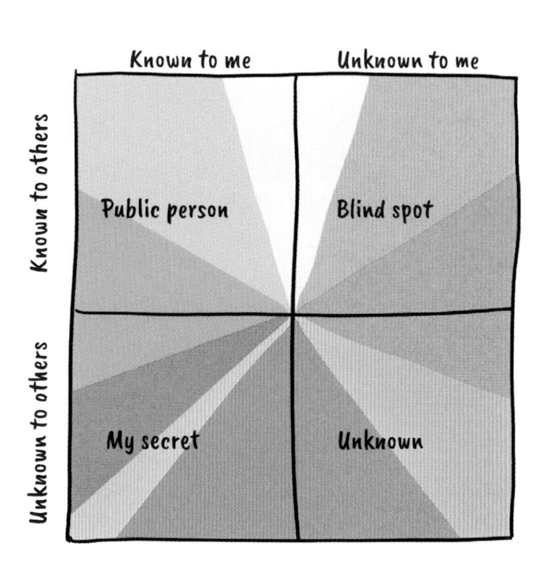

In terms of career and studies, for example, such a crisis manifests itself in the form of burnout, or it can be witnessed when our exam grades get worse and worse. It is therefore important that we constantly reflect on and question the hypotheses on which our actions are based. We regularly check their usefulness because hypotheses in human systems have to be re-created over and over again. The reason for this is that the context and the framework in which we operate are constantly changing.

If we don't throw our hypotheses overboard, we are stuck on the hamster wheel until we arrive at resignation or burnout.

But questioning hypotheses is difficult when they affect your own life because this requires constantly questioning assumptions that have been useful so far. As a result, you have to give up routines and thought patterns that you have grown fond of over the years. The time has come to make room for something new. For Steve, for example, the hypotheses that have been valid so far were strongly influenced by his family. Only through his studies was he able to make his "own way of thinking" the object of reflection. Now the time had come to formulate new hypotheses.

Formulate hypotheses

The **first step** toward devising a new hypothesis is reflection. The result of your first reflections on your values, thinking preferences, and motivating environmental factors will already give you initial clues for what is important to you. You can also reflect on various aspects. For example, you may ask yourself:

Which subjects did you particularly enjoy at school, in training, and during your studies?

What topics have you been interested in lately when you talked to colleagues and friends?

Which topic would you spend your free time on to deepen and learn more about it?

Are there services or products that fascinate you? Why are they of interest to you, and would you like to know more about how these services are created and who is behind them?

The **second step** is to formulate two to three statements based on your interests and choose the one you like the most from among them. For example, Steve might write:

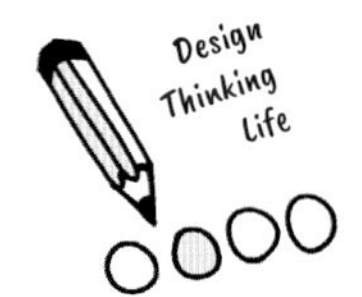

I like economic subjects in general and innovations. I'm particularly interested in new technologies and how they can be incorporated profitably into the future of work.

The **third step** is to formulate a variety of "I believe, I want to..." statements. Steve has formulated the following sentence, among many other statements:

I think I want to work in a company that uses cutting-edge technology or is developing it in-house.

The **fourth step** is the comparison with your current situation, the insights just gained, and the previously prevailing hypotheses.

How does your "new worldview" fit into the existing system?
What is the divide between your "personal world" and the new circumstances?

Result of the reflection

The reflections we have undertaken on our skills, values, and motivating environmental factors help us select opportunities and options for our professional and career planning.

Steve's skills, values, and motivating environmental factors

- Steve's thinking preference is on the right side of his brain. For example, he can deal with other people very well and is very empathetic. He also solves problems intuitively.

Steve's HBDI profile

- In terms of values, Steve puts "independence," "standard of living," "activity," and "belonging" at the top of the pyramid.
- When reflecting on the environmental factors, Steve attaches particular importance to "safety and stability," "clear rules," "friendly, open atmosphere," and "little detail work."

Find your "BIG LIKE," and perceive the possibilities that are offered.

What new insights has your reflection yielded in terms of your skills, values, environmental factors, and comparison of existing and new hypotheses?

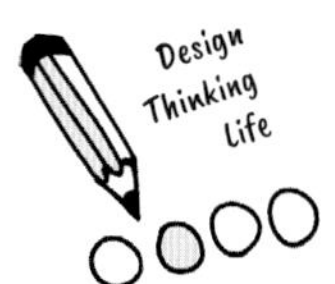

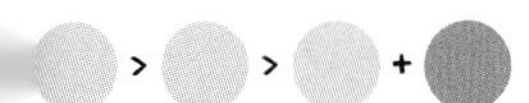

Design career paths

Each of us is different, so our professional and career paths are different as well. As the saying goes, "Many roads lead to Rome," and, after all, a career is always successful when we feel happy with what we do. When designing career paths, we want to imagine, test, and reflect on as many paths as possible. This iteration helps us make a better selection of options and thus either create the space for change or confirm what we are currently doing.

Create options for professional and career planning

From our experience, there are usually two critical situations in terms of career planning. Either we are in a vacuum and generally do not know what to do in terms of a career or a profession, or we are one step ahead, options are open to us, but we don't know which one we should choose. If you have tangible options, you can go directly to page 226. However, we recommend that you go through this exercise in either case because it often yields new and unexpected insights.

Do not create only one career plan but consider many different versions. Even if you have already embarked on a path, different options are valuable for your development.

For the first variant, i.e., when we don't have a clue as to what should come next, we design three professional and career plans – namely, with the following characteristics:

1. Continue your current career with which you are more or less satisfied: "You follow the path that you are already pursuing."

2. Design a new career, which means that the current position no longer exists: "You need to come up with a plan B."

3. Design a life in which money and status play no role: "You can do what you enjoy doing."

Steve finds the following three focal points exciting: "Continue as before"; "What would happen if the current path no longer existed"; and the "Luna Park" variant, i.e., do only what you like doing. Based on the three focal points, he creates three life plans. His bachelor's degree in management science and engineering always constitutes the starting point.

Steve's life plan: "Go on studying"

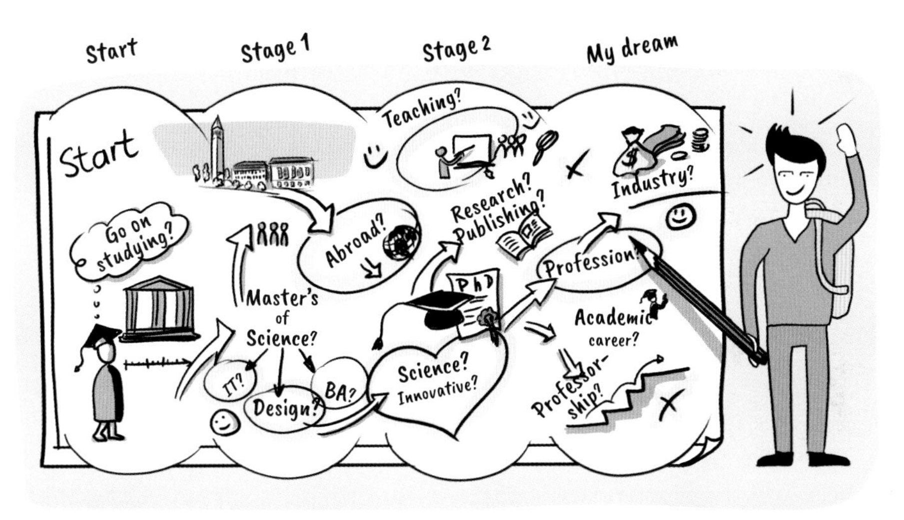

What needs does the life plan address?

- Is research something I can imagine?
- Should I continue studying business information systems or something else?
- Do I want to work in the industrial sector or at a university?

What questions does the life plan leave open?

- An MSc is a good idea, but IT is not really my thing.
- A new environment and a new country would tempt me.
- I cannot currently imagine what working at a university would be like.

Holistic view of the life plan

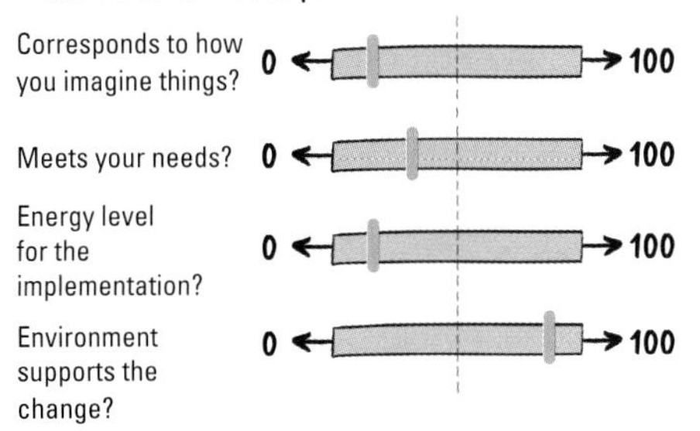

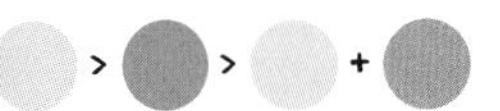

Steve's life plan: "Studying is not an option"

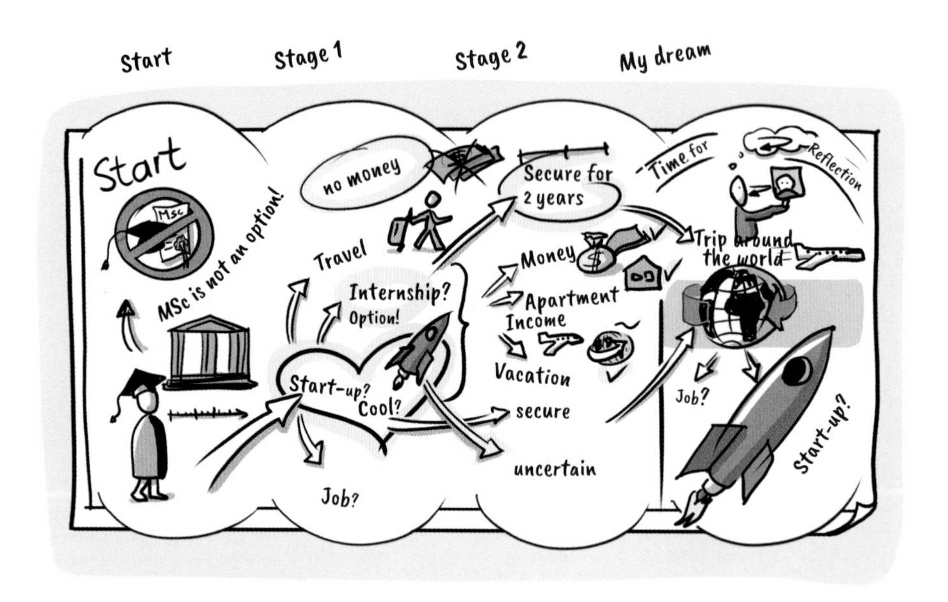

What needs does the life plan address?

- Can I create added value without any experience in a start-up?
- Might a trainee program be a good alternative?
- Will all paths be open to me afterward?

What questions does the life plan leave open?

- Working at a start-up sounds "cool," but it's uncertain.
- An internship would give me the opportunity to get to know different areas at a company; and it takes one year max.
- Somehow, the big, wide world lures me, and a longer trip is my dream. But for this, I have to save some money first.

Holistic view of the life plan

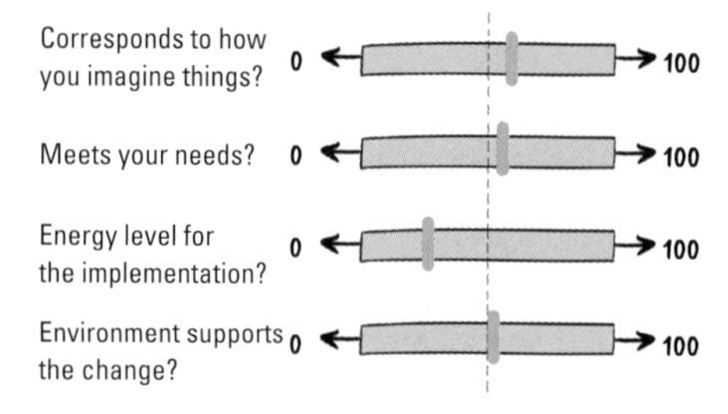

Steve's life plan: "Only do what you enjoy doing"

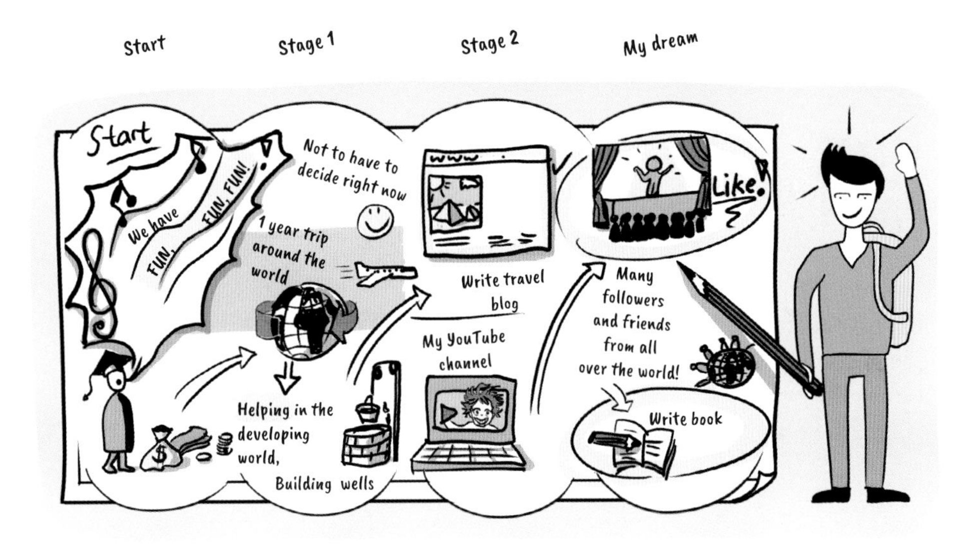

What needs does the life plan address?

- Do I even want to travel around the world that long?
- How do I find projects I can help with?
- Does becoming famous correspond to how I imagine things and my personality?

What questions does the life plan leave open?

- Just leaving everything behind now and taking a trip around the world sounds great, but I'm not sure if I'd dare to do it right away.
- The idea of helping somehow and finding creative solutions is a beautiful notion that makes me feel good.
- Actually, when I think about it, I don't want to become famous. At most, I want to show my ex-girlfriend that I "have a lot going for me."

Holistic view of the life plan

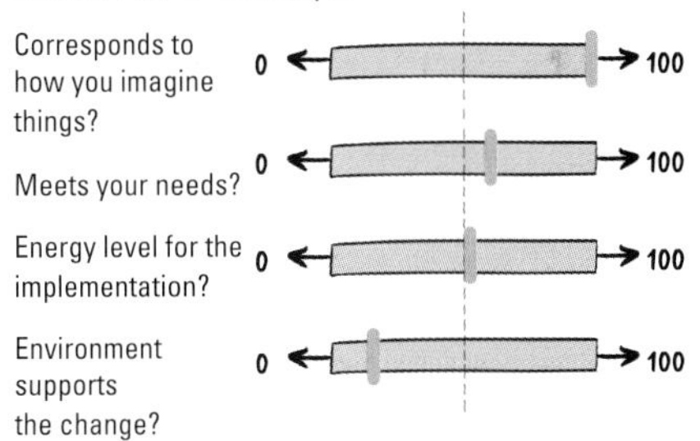

How would your career go if you didn't change anything?

Career plan 1: . . . .

Start

Stage 1

Stage 2

What needs does the career plan address?

. . . .

. . . .

. . . .

What questions does the career plan leave open?

. . . .

. . . .

. . . .

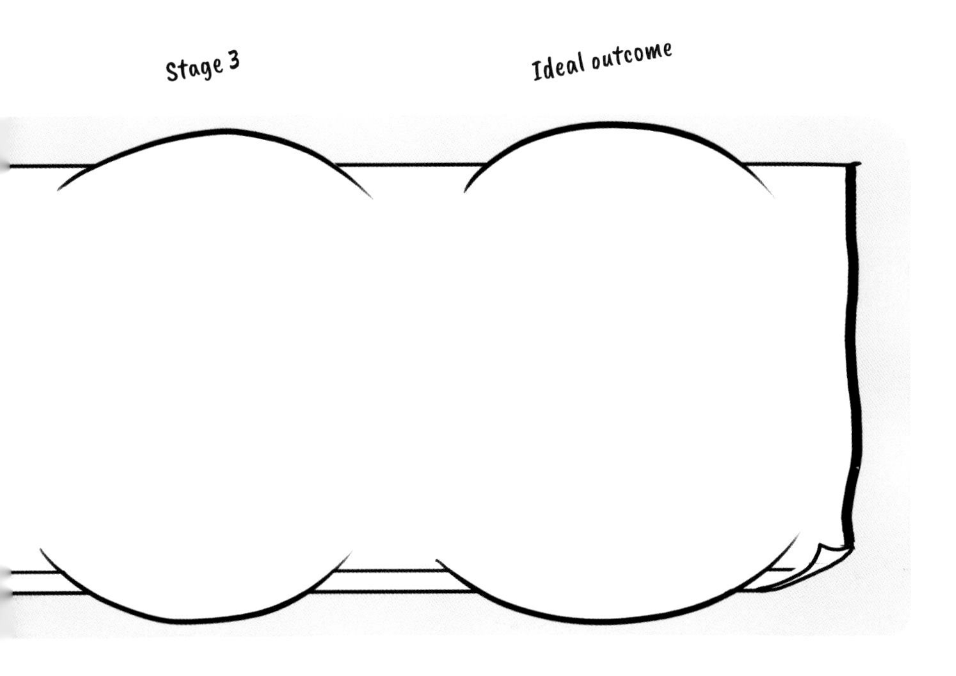

Holistic view of the career plan

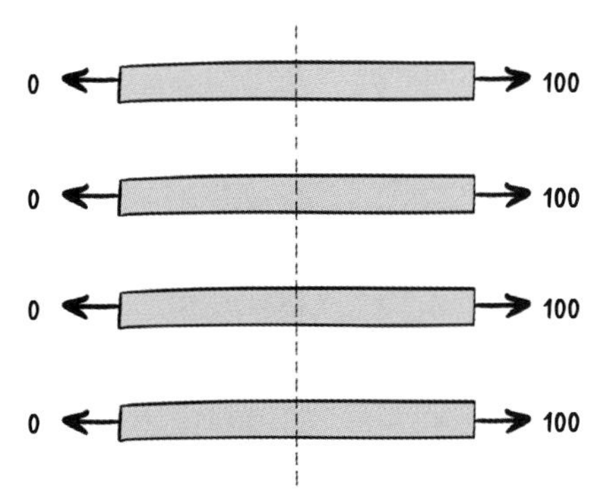

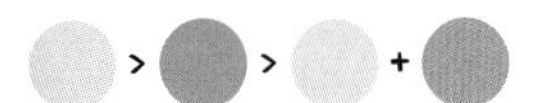

What would you do if the current position no longer existed?

Career plan 2: . . . .

Start

Stage 1

Stage 2

What needs does the career plan address?

. . . .

. . . .

. . . .

What questions does the career plan leave open?

. . . .

. . . .

. . . .

PLAN B

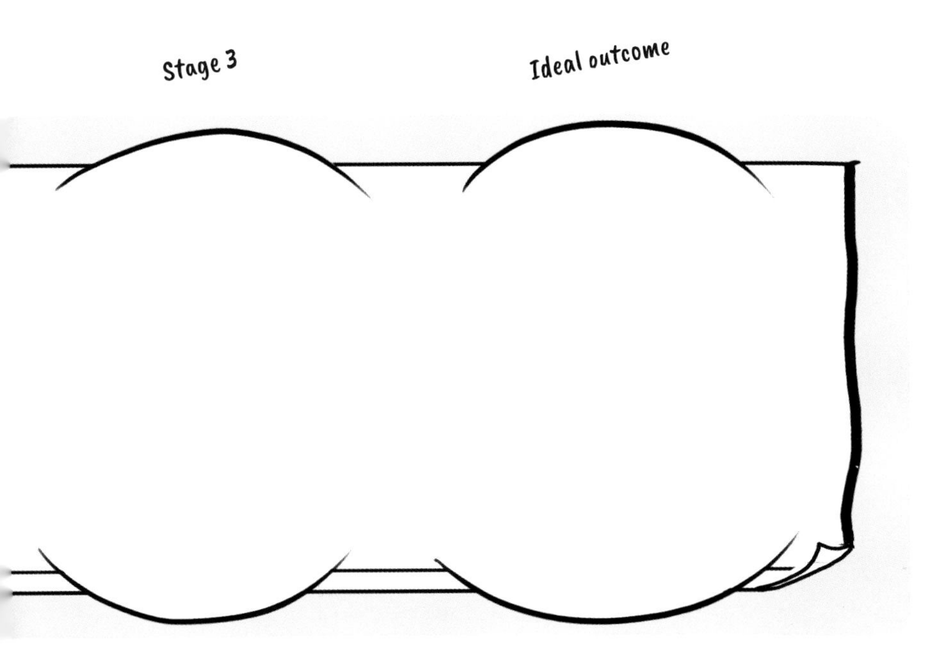
Stage 3
Ideal outcome

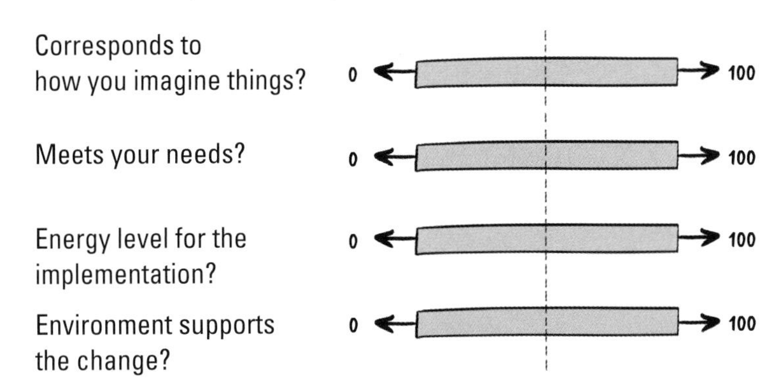
Holistic view of the career plan
Corresponds to how you imagine things?
0
100
Meets your needs?
0
100
Energy level for the implementation?
0
100
Environment supports the change?
0
100

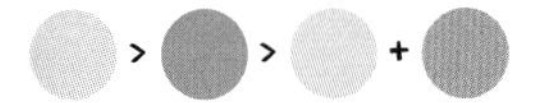

What would your career or situation look like if money and status didn't matter?

Career plan 3: . . . .

Start

Stage 1

Stage 2

What needs does the career plan address?

. . . .

. . . .

. . . .

What questions does the career plan leave open?

. . . .

. . . .

. . . .

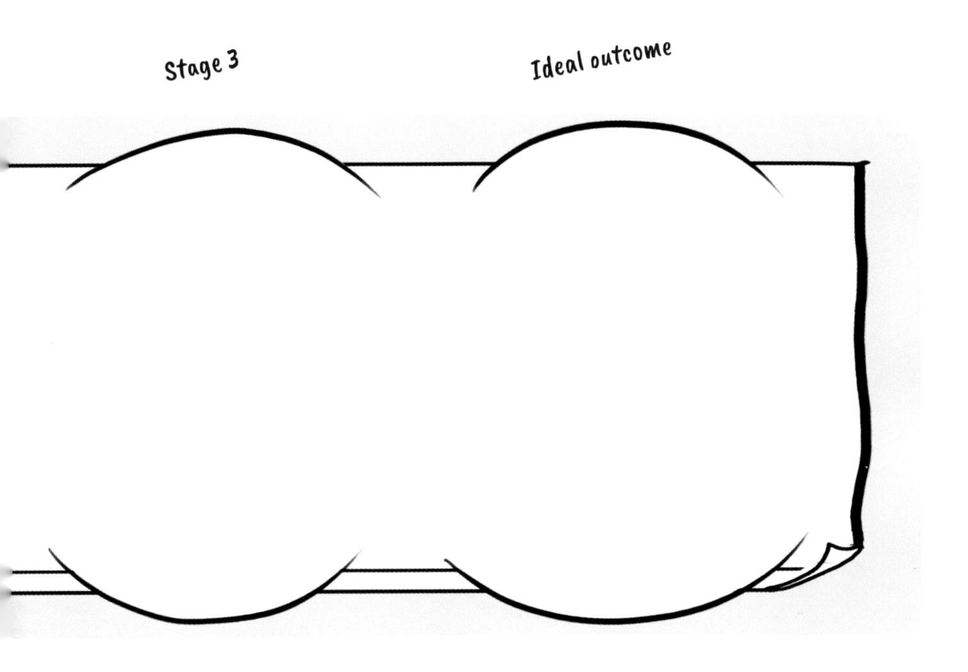

Holistic view of the career plan

Corresponds to how you imagine things?	0 ← → 100
Meets your needs?	0 ← → 100
Energy level for the implementation?	0 ← → 100
Environment supports the change?	0 ← → 100

Evaluate, test, and implement options

Making the right choice is easier said than done. And although we now know there are always and at all times several possibilities, we find it difficult to go in a new direction and leave our existing paths. The reason is that we have settled in our comfort zone or often think that our daily hamster wheel will eventually take us to a point where we'll feel joy. If we have gained new insights in the previous DTL process, now is the right time to initiate a change.

Focus on which option?

In design thinking, suitable solutions evolve where these three things intersect: the user's needs (desirability), a solution that is profitable (viable), and feasibility.

Three central questions can be derived from this for professional and career planning:

1. What would you like to do?
2. Which skills do you have?
3. What kind of talent is the market looking for?

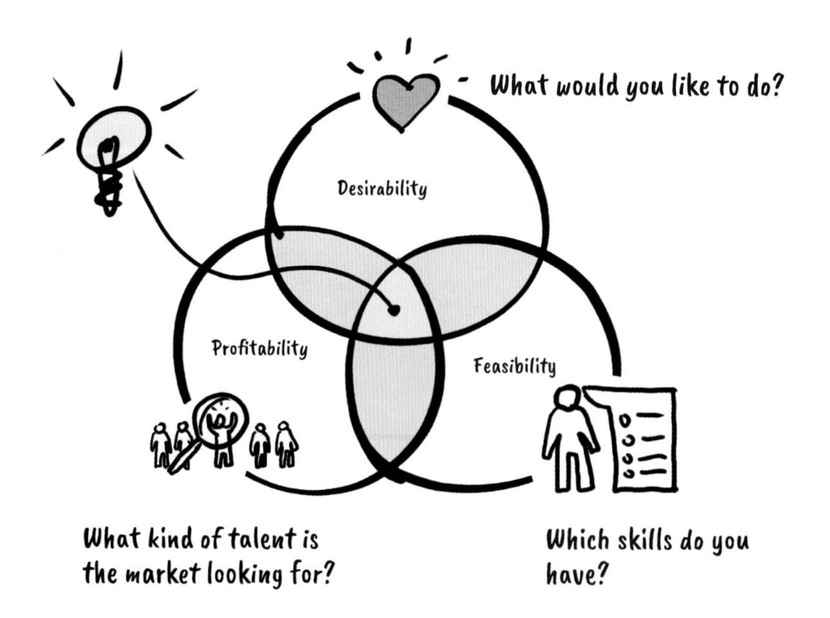

Why are the three dimensions important?

- They reduce your risk of making a wrong career move.
- They help you find out more quickly what you really want to do.
- They open up new perspectives and opportunities.

So far, we have ignored profitability in designing career plans and have focused mainly on desirability and feasibility.

The latter two properties are of prime importance to us in the "Design Thinking Life," because we're convinced that we can be successful only if we pursue an activity that gives us pleasure. So our definition of a successful career is not based on annual income but instead on doing something that we are proud of, we are recognized for, and we feel is meaningful. In other words, these are answers to the question: What are we doing something for?

Nevertheless, profitability is part of professional and career planning, and we should analyze which industries, companies, and tasks will continue to exist in the future. Such forecasts can be taken from statistics, foresight, and trend papers. In addition, there are professions that will disappear sooner or later as a result of increasing digitization and automation. However, there are also innumerable new fields of activity that arise as a result and offer new opportunities. Often, the content and the skills required in a particular profession change as well.

So if you want to stay fit for the job market, you have to keep on training to meet the changing requirements.

We would like to take this opportunity to remind you that all areas of life are interrelated. In the first part of the *DTL Playbook*, we presented this system as a mobile. Profession and career are part of this system, and only if it is balanced can we lead a fulfilled life. On the following pages, we have provided techniques and strategies to simplify the selection of options.

In the DESIGN THINKING LIFE with regard to career, it is about your needs, your skills, and matching your talent with the demands of the labor market.

Selection of ideas and concepts

Many ideas and concepts are wonderful, but choosing one is often difficult, given their abundance. To reflect on your three career plans in terms of desirability, feasibility, and profitability, the following questions may help you.

Career plan	1	2	3
1) Desirability: Which plan comes off best on the self-reflection dashboard?			
2) Feasibility: Which plan matches your skills, values, and environmental factors best?			
3) Profitability: What is the most promising plan in terms of a job and adequate pay?			
Which plans or parts of plans are exciting, seem cool, and should be tested?			

Notes:

The tetralemma

In the evaluation of career plans, we've noticed that it is not so easy to choose between two options. For example, in the case of a course of study, the question of whether to study business administration or IT; or in the context of the next career step, whether to choose having more money or more free time.

For example, when formulating options for professional and career planning on page 216 et seq., we continued the current path in option one; designed a completely new career in option two; and with option three, we could simply do what we enjoy doing.

Through our subsequent evaluation, we tried to adopt an either-or attitude. However, the tetralemma method goes one step further and looks specifically for "as well as" possibilities and new considerations, if no option is convincing at all. Based on our experience, there will often be – based on the three defined options – "as well as" solutions. For example, the "more free time or more money" dilemma might turn into a job with the same income but a 90% workload.

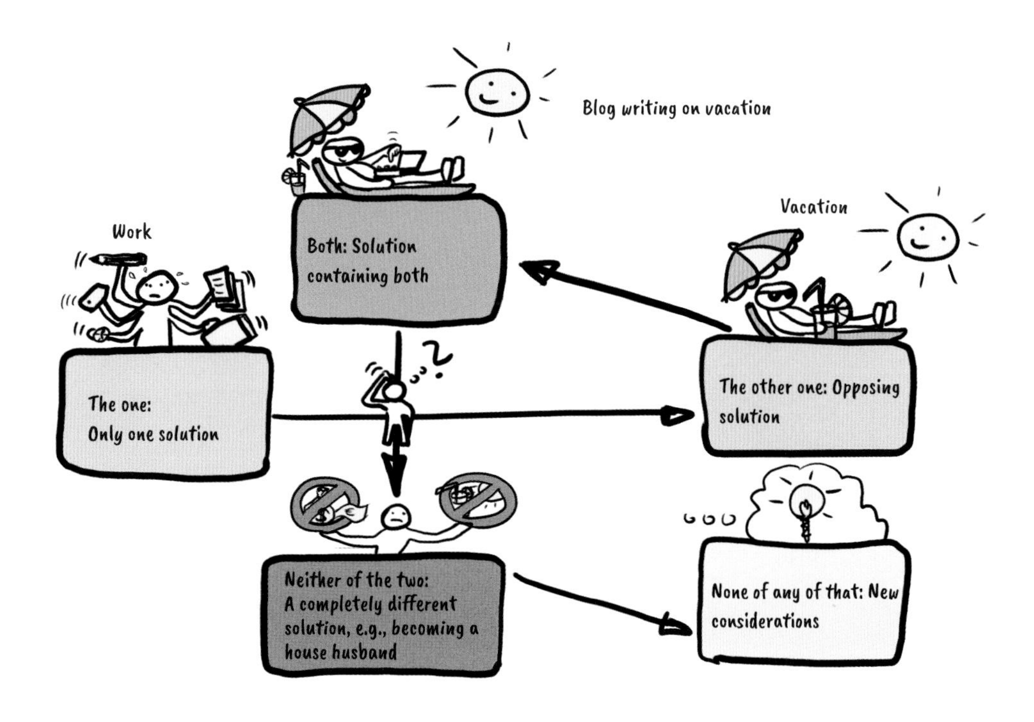

How can we test the defined and combined career plans?

To test the possibilities and the ideas, we showed various procedures on page 158 et seq. It often makes sense to use brainstorming or brainwriting for this step, in order to test the preferred career plans in the right context. Sometimes family members and friends also have good ideas and can help with brainstorming. So why not get support?

Steve's brainstorming

Steve's first brainstorming sessions focus on potential opportunities to test his career plan and potential companies, organizations, and industries that fit his career plan. After brainstorming, he then selects the best ideas and creates a list of actions that are necessary for him to explore the opportunities.

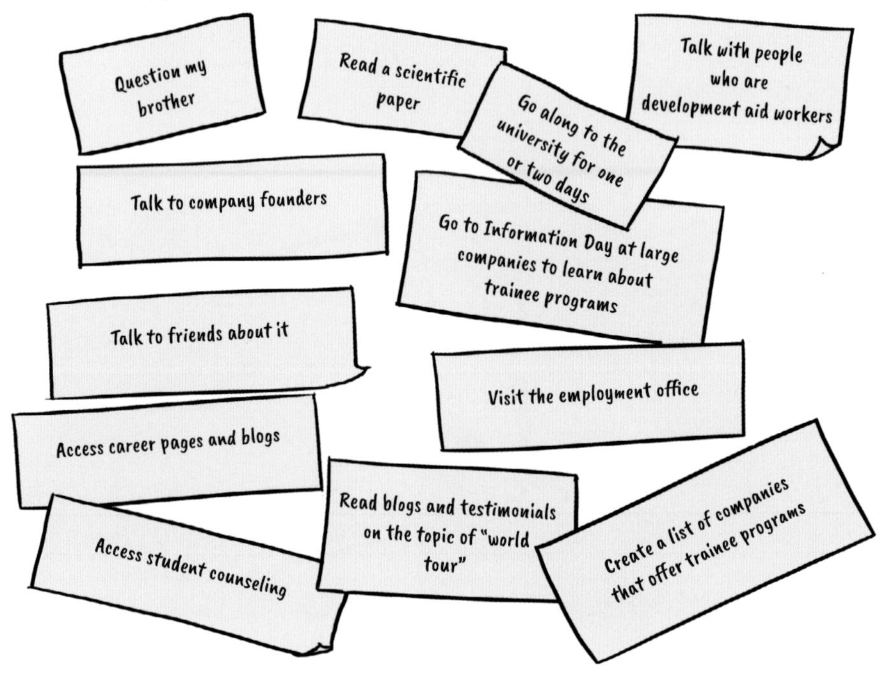

Who? What? Where? When?

Steve makes a list of the activities.
He becomes as specific as possible in his planning.

1. Talk to my brother and accompany him to Cornell University for a day or two, preferably next week when he is visiting the family in New York for a family celebration.
2. Trial work in a technology start-up, ...
3. ......
4. ......

Brainstorming to test your career plan

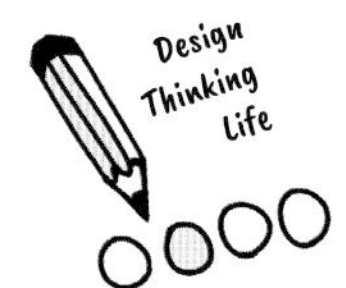

What possibilities are there to test your career plan?
Which positions, companies, organizations, and industries match the career planning?
At the end, create a list to determine with whom you test what and when you do it.

Brainstorming to test your career plan

Make a list with activities
(Who? What? When? Where?)

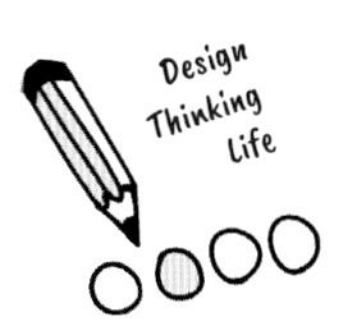

WHO	WHAT	WHEN	WHERE	
				☐
				☐
				☐
				☐
				☐

Test options

So far, like Steve, we have evaluated our options only on the basis of our skills and values with various tools.

Now it's time to test our ideas to see how we feel and find our way to the situations we want to live in.

We experiment with our career plans to get answers to our questions, to collect experience, and finally to test our hypotheses.

Steve's approach to test the options of a doctorate, a start-up, or an internship

Situation description by Steve:

I spent two days with my brother at the university. One day, there was an internal conference where topics were discussed in small break-out sessions. I enjoyed the discussions enormously and appreciated the interactions among colleagues at the university.

On the second day, I experienced what an average working day is like for my brother. He held a tutorial, he gave a lecture, and in the afternoon there were very boring meetings with various expert committees. This day seemed like half an eternity to me, and I was very happy when I could leave the university with my brother at 6:00 p.m.

HIS CONCLUSION: RATHER BORING

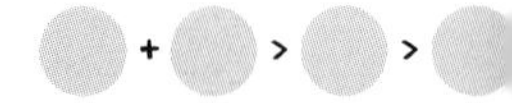

I also had the opportunity to work in a start-up for two days. On the first day, I was allowed to meet the CEO, but he didn't have much time. He gave me a brief introduction and access to a software repository. Then I was on my own. Colleagues had little time to answer my questions, and when I asked for guidelines, I realized that everything was a bit chaotic.

HIS CONCLUSION: TOO LITTLE STRUCTURE

The first day that Steve had spent with his brother was very exciting. The second day, he was bored to death. The interaction with company founders was also rather sobering for Steve. There was a lot to do, and the company had little structure. Usually, Steve was quite alone with his problems and questions.

Situation description by Steve:

At a birthday party with friends, I told them about DTL and that I was thinking of either continuing my studies or working. This situation looked familiar to a lady friend of mine. She told me about her trainee program.

HIS CONCLUSION: COOL OPTION

Steve's conversation with a friend of his gave him another interesting perspective on starting a career. She had done a twelve-month internship at Apple in Santa Clara. The Software Engineering Internship Program allowed her to apply and experience her critical thinking skills, her problem-solving skills, and her ability to work quickly and purposefully with teams-of-teams. Through interdisciplinary collaboration with many other departments, she also experienced the work of designers, marketing experts, and product managers.

We would like to test our options to be able to gain as many new insights as possible. Only in this way is it possible for us to learn more about our desired future. In addition, self-reflection is of vital importance on our "journey."

Collect and analyze insights

Through shadowing, interviews, internships, and small explorative projects, you can gain insights that you should document. Typical questions:

How did it feel?
What made you happy or irritated you?
Which career plans or combinations of different plans do you prefer?
What have others done in this situation?

Design
Thinking
Life

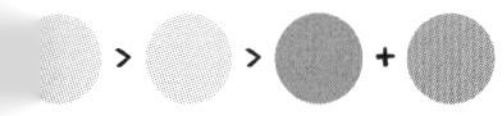

Focus and implement

It's time to make a choice about our career or individual stages of it and take the first steps of implementation.

We take a step back and consider, from a distance, our self-reflection, options, and results from our experiments. This is the best way to see which solution has turned out to be the right one.

For the implementation, we recommend that you take some time, initiate the first small steps, and tackle them one by one. A vision, a plan, and a to-do list are effective tools for navigation, but much more important is simply to get started. Based on our experience, it's best to tackle small stages, celebrate successes, and thus initiate change gradually.

In addition, we often want to complete too many tasks immediately and at the same time. We run off without knowing where to go. Let's give all credit to agility and speed where it's due, but thinking, acting, and reflecting are three important principles for the successful application of this "DESIGN THINKING LIFE" mindset.

From the inside, a hamster wheel looks like a career ladder.

First stage of a new path

In the meantime, you are practiced in directing your energy toward positive things. This is why you should now put a lot of energy and commitment into the first stage of a new path. And at this point, you need no more new methods from us. You'll simply act because you'll take the time to do so and enjoy it. And you'll regularly question your hypotheses.

- What do you do first?
- Who can help you?
- By when would you like to implement this stage?

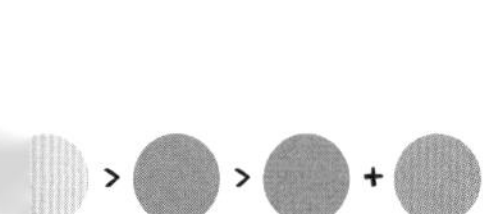

Finally, the question: What happened to Steve?

Steve had to make bigger decisions. After getting his bachelor's degree, he has several options. On one hand, he could graduate with a master's degree and, like his brother, earn a doctorate, although his observations rather discourage him from following this path. His brother lived with his wife, Maren, and their little daughter in the attic of their parents' house during the time he was getting his doctorate. Money was tight, and he often worked many hours at the university to prepare lectures, in addition to his research work. Maren and Alex have reconsidered their plans and are preparing to move to Delaware, which stands for "liberty and independence." The port of Wilmington is not as impressive as that of Singapore, but the area has a lot of unspoiled nature nearby and great shopping for Maren in Philadelphia, only forty-five minutes away. During the two days he spent with his brother, Steve had positive experiences, but academia has very boring elements as well. The start-up idea sounds very exciting to him, but Steve has reservations about what he would be able to contribute. So far, he has no work experience, except for a summer job as the coordinator at a Boy Scout camp.

Mental overload

In conversations with company founders, he found out that many of them are between thirty and forty years old and often have already failed with one or two plans before they founded a successful company.

A friend of Steve's came up with the idea of taking part in a twelve-month internship program with one of Silicon Valley's tech giants. In one of the internship programs offered, Steve would also have the opportunity to get a taste of up to three different divisions during the twelve months. Steve is particularly attracted to this offer because it would give him the opportunity to get an insight into different divisions of a company and see what he really likes. In addition, he'll gain some time with regard to the idea of pursuing a master's degree. After the internship, everything will be open to him: go on studying, join a start-up, or enter professional life.

When Steve started working with DTL, he was a little disoriented. It was important to him to find out where his strengths and weaknesses, preferences for action, and thinking, as well as his inclinations, lie. In the next year, he will test these new assumptions during his internship!

Questions for reflection on options

Major changes have an impact. And in order to leave the hamster wheel, we often have to adapt structures, processes, and rules radically. At the end of Part II, we will give you eight questions for reflection on your individually chosen options, the answers to which will quickly show whether you are ready to work not only *in* the system but also *on* it. If you still have doubts about your preferred option, it makes sense to think again about individual statements or to rethink your career path.

Decision for reflecting on an option

1) You are aware of the goal you want to achieve with the change and the areas of life the change will impact.

2) You reflect on what contributed to the success or failure of the last big changes, and you use these insights for your current situation.

3) You have become aware of who will be affected by the new situation. You know the proponents of change and who might have something against it.

4) You know what you have to do without and what to let go of when you have chosen an option. You're ready to accept this sacrifice, which comes with every change.

5) You know what you need to do next to achieve your goal.

6) You know who will support you in your project, and you have identified people who initiated similar changes and can help you with advice.

7) You have imagined in your mind how you will think about the upcoming change in two or five years when looking back.

8) After you have examined the chosen option with the answers arising from this reflection, are you sticking to your decision?

Last but not least – The journey's end is really the beginning

The DTL mindset helps us with our own change, but it is also the basis for changing organizations, companies, and other systems. Self-efficacy and self-awareness are the keys not only to dreaming but to taking your life into your own hands and actively living it according to your own ideas. At the end of this playbook, we would like to encourage you once again to see the design of change as something continuous, ultimately to achieve more well-being. The rules of thought summarize the most important DTL techniques and strategies and make us aware that we write the script of lives ourselves and that we can always – at every given moment – rewrite it, i.e., change it.

We have arrived at the last pages of the DTL book. Actually, we might as well start again on page 1 because our life, in a spiral development, is permanently in motion, and new questions will arise again and again in the different areas of life. In addition, if we reflect at regular intervals on where we stand at the moment, it has a positive impact on our well-being and self-efficacy.

Through DTL, we have realized that there are situations in life that we have to accept and solvable problems we can actively address. Sometimes, it also helps to look at problems from different angles (reframing), in order to deal with them differently and in a better way or simply to be able to accept them. A deep understanding of ourselves and others' views of our actions give us important stimuli to make changes. What's important here is our mindset (our attitude), combined with strategies and techniques that accompany our changes.

In the "DESIGN THINKING LIFE" paradigm, it's essential to test our prototypes and concepts, as well as perform iterations, so we are finally able to break new ground. The ability to leave familiar shores can be learned. If we apply this concept to ourselves, we are well prepared to use this mindset profitably in our daily work, in the transformation of teams, or in organizations and companies. People who succeed in doing so write the script of their lives themselves in a self-efficacious way!

The following rules of thought are helpful to motivate ourselves daily to design change, to stay on the path to self-efficacy, and to increase our well-being:

- Think the unthinkable – **"think outside the box."**
- **Distinguish** between **facts** and **solvable problems.** A good problem definition is the best starting point for designing change.
- **Reflect on** statements from **others' perceptions of you**. They are indications that you might have to reject your hypotheses and reorient yourself.
- Change always starts **with yourself**. Making others responsible for your situation is good for your mental hygiene in the short term, but it does not enable you to act.
- **Test new life concepts** and dare to push yourself to the limits – only then can you exceed these limits at some point.
- Have a **big vision,** which you will implement or adapt in **small steps**.
- Change always entails **resistance**. Knowing who supports and who blocks you helps you define the next steps.
- And last but not least: Everything usually transpires differently than you thought it would, and that is why the design of your life is a **continuous task**, in which new directions can be taken at any time.

Only those who accept their yesterday and today can design their tomorrow freely.

Only those who let go have free hands to seize the future.

Index

A

Acceptance phase, 44
Activity/energy chart, 62–63
AEIOU-Fragen, 64, 72, 82
"Alex," 187–188
Analogies, 129, 132
Authenticity, 91–92

B

"Back to the Future" (warm-up), 20–23
Big change, initiating, 186–187
Brainwriting, 116

C

Career exploration framework, 196–197
Carleton, Tamara, 124
Change, big vs. small, 176–177
Cockayne, Bill, 124
Context map, 100
Converging, 116, 138
Csikszentmihalyi, Mihaly, 54

D

Dark horse, 138
De Shazer, Steve, 170
Design thinking, defined, 19
Design thinking life (DTL), defined, 28
Design thinking mindset, 19, 26, 27
Diverging, 116
Double loops, 176
DTL (design thinking life), defined, 28

E

Ebner-Eschenbach, Marie von, 177
Energy journal, 55–58
Environmental factors, 204
Epictetus, 47
Exercises:
- AEIOU tool, 72, 82
- "Back to the Future" (warm-up), 20–23
- brainstorming, 233
- brainstorming with analogies, 130–131, 133–135
- career planning, 220–225
- context map, 102–105
- define stages and goals, 114–115
- desire pyramid, 193
- energy journal, 66–87
- facts vs. solvable problems, 45
- feedback capture grid, 163–167
- First stage planning, 241
- Formulate hypotheses, 208–211
- HBDI model, 200–201
- idea selection, 126–128, 136–137, 142–144, 230
- Ideate, 119–123
- network mapping, 169
- others' perception of us, 95–97
- point of view, 108–109
- rating of favorites, 145
- reframing, 48–49
- self-check, 180–183

Exercises: (*continued*)
self-dialogue, 50–53
self-reflection, 37–39, 171
Sketch plans and stages, 150–156
taking stock, 30–31
10-point challenge, 47–48
values pyramid, 203

F

Failure, 158
Feedback capture grid, 160–161
Flow, 54–55
Focus, 240

G

Groan zone, 138–139

H

HBDI (Herrmann Brain Dominance Instrument) model, 198
Hypothesis, overcoming, 206–207

I

Idea selection (daisy), 124
Idea selection questions, 246–247
Ideation, 112–113
analogies, 129
brainwriting, 116
groan zone, 138–139
idea selection (daisy), 124
summarize and select ideas, 140
Implementation, 170, 240
Internships, 158

J

"John":
analogies, 130
brainwriting, 116–117
energy journal, 60–61
life plan, 162
persona description, 16, 59–60, 172, 174
point of view, 107
Journal, energy, 55–58

M

"Maren," 187–188

N

Network, importance of, 168

O

Optimists, 46
Others' perception of us, 93

P

Personal warning system, 54–55
Pessimists, 46
Picasso, Pablo, 32
Point of view (PoV), 100, 106
Professional and career planning, 189, 216, 228
Profitability, 229
Prototypes, 148

Q

Quarter-life crisis, 191

R

Reframing, 46–48, 170

S

Self-check, 180
Self-dialogue, 50–53

Self-efficacy, 49, 190–191
Self-image, 93
Self-reflection, 30–39, 171, 198–205, 208
Shadowing, 158
Single loops, 176
Social media, 91–92
Socrates, 93
Stanford University, 124
"Steve":
  brainstorming, 232–233
  HBDI model, 199, 212
  life plan, 217–219
  persona description, 17, 187, 194–195, 242–243
  testing, 236–237
"Sue":
  analogies, 129, 132
  brainwriting, 118
  context map, 101
  idea selection, 125, 144
  ideation, 113
  others' perception of, 94
  persona description, 17, 34–35, 172–173
  point of view, 106
  prototyping, 148–149
  selection of ideas, 140–141
  self-reflection, 36
  testing prototypes, 160
Suzuki, Shunryu, 18

T

Teamwork, 168
10-point challenge, 47–48
Testing, 157–158, 232
Tetralemma method, 231
Thinking preferences, 199

V

Values, 202
Visualization, 32

W

WH questions, 95

DESIGN THINKING LIFE applied

The DESIGN THINKING LIFE mindset helps people in different phases of their lives. At the end of the book, we would like to inspire you with selected pictures from DTL workshops. Experiencing DTL as an online course or with others can be very enriching.

Example:
DESIGN THINKING LIFE workshops for students and professionals

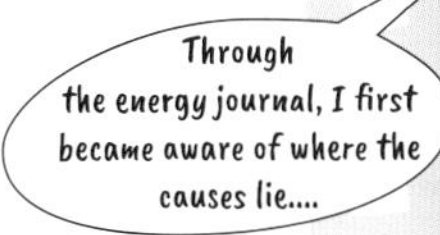

Example:
DESIGN THINKING LIFE/
DESIGN YOUR FUTURE
online course

Example:
DESIGN THINKING LIFE
workshop for career planning

Example:
DESIGN THINKING LIFE
training for professionals, managers, and executives

Want to learn more about design thinking?

The Design Thinking Playbook
Michael Lewrick, Patrick Link, and Larry Leifer (eds.)
With traditional, current, and future factors of success
ISBN 978-1-11946-747-2
352 pages · paperback

MINDFUL DIGITAL TRANSFORMATION OF TEAMS, PRODUCTS, SERVICES, BUSINESSES, AND ECOSYSTEMS

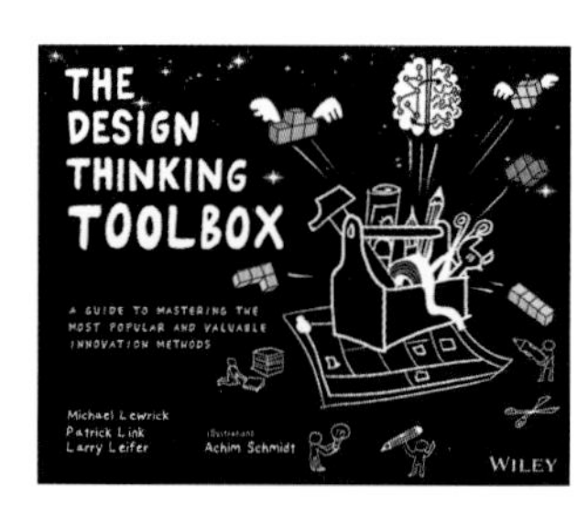

The Design Thinking Toolbox
Michael Lewrick, Patrick Link, and Larry Leifer (eds.)
Tips and tricks from the Design Thinking Community
ISBN 978-1-11962-919-1
320 pages · paperback

A GUIDE TO MASTERING THE MOST POPULAR AND VALUABLE INNOVATION METHODS